HEAT CONDUCTION

**International Series
in Heat and Mass Transfer**

Arthur E. Bergles and Ulrich Grigull, *Editors*

Grigull and Sandner, Heat Conduction (translated by J. Kestin)

HEAT CONDUCTION

Ulrich Grigull

Heinrich Sandner

Technical University of Munich
Federal Republic of Germany

Translated by
Joseph Kestin

Brown University
Providence, Rhode Island

SPRINGER–VERLAG

Berlin   Heidelberg   New York   Tokyo

DISTRIBUTION IN NORTH AMERICA

HEMISPHERE PUBLISHING CORPORATION

Washington   New York   London

HEAT CONDUCTION

1 2 3 4 5 6 7 8 9 0   E B E B   8 9 8 7 6 5 4

This book was set in Press Roman by Hemisphere Publishing Corporation. The
editor was Brenda Munz Brienza; the production supervisor was Miriam Gonzalez;
and the typesetter was Sandra F. Watts.
Edwards Brothers, Inc. was printer and binder.

Library of Congress Cataloging in Publication Data

Grigull, Ulrich, date
  Heat conduction.

  (International series in heat and mass transfer)
  Translation of: Wärmeleitung
  Bibliography: p.
  Includes index.
  1. Heat—Conduction. I. Sandner, Heinrich.
II. Title. III. Series.
TJ260.G7513 1984   621.402′2   83-16616

ISBN-13: 978-3-642-96818-1   e-ISBN-13: 978-3-642-96816-7
DOI: 10.1007/978-3-642-96816-7

DISTRIBUTION OUTSIDE NORTH AMERICA:

CONTENTS

LIST OF TABLES

PREFACE

The present book treats the science of heat conduction to an extent to which it can be taught in the specialized departments of Mechanical, Chemical or Electrical Engineering at a German Engineering University. No special prerequisites are assumed, and the mathematical methods employed draw, essentially, on the content of a normal curriculum in the departments mentioned above. The book is intended for adoption in conjunction with a standard lecture course or as an aid to review before examinations. It should also be found helpful to a practicing engineer in solving problems in heat conduction.

As far as the readers are concerned, the book should, above all, show that exact and approximate solutions to answer questions which arise in a very large number of important, practical applications are at their disposal. Further, the book will show that in many cases it is possible to perform first estimates in a very elementary manner before engaging in the derivation of complicated analytic solutions. It is in this way that we utilize the past results of the great mathematicians of earlier generations who have bequeathed to us a considerable stock of methods and solutions. The application of such methods is illustrated in this book with the aid of examples drawn from various branches of science and technology. In this manner, the wide field of applicability of heat transfer will be made clear.

Often, the application of theory is made difficult because the required thermophysical properties are not known, especially the values of thermal conductivity and thermal diffusivity. For this reason, the book contains a large number of numerical tables which, to a certain extent, include the effect of temperature on properties. A separate chapter is devoted to an introduction to the theory of transport properties.

As a matter of principle, the equations in this book are written in the form of quantity equations; exceptions are specifically indicated. In the treatment of numerical examples and in the tables, we have employed the International System of units, or SI, together with the recognized multiples and submultiples. Conversion tables facilitate the use of older references.

This book provides a continuation of *Grundgesetze der Wärme-übertragung* by H. Gröber and S. Erk whose third edition was at one time prepared by the senior author of this book.[†] The preparation of additional volumes is envisaged.

U. Grigull
H. Sandner

[†] The English translation is entitled *Fundamentals of Heat Transfer* by H. Gröber and S. Erk, 3d ed. revised by Dr. Ulrich Grigull, translated by Dr. Jerzy Moszynski, McGraw-Hill, 1961.

TRANSLATOR'S PREFACE

Not much new can be said about conduction in an elementary way. The merit of the present book is that it teaches the subject to a novice in a simple and direct manner. In addition to its great clarity, the book is characterized by a felicitous selection of topics; felicitous that is, both in what it contains and in what it omits. The emphasis is not only on simple analytic methods, even though these are covered more than adequately, but on the development of an intuitive "feeling" for numerical values, so important for success in design. The cumulative effect of the exposition and of the worked examples is to implant in the student-reader a visual appreciation of rates of heat transferred by conduction, of sizes, of the effects of varying and changing properties, and, above all, of the importance of thinking in numbers and not only in symbols.

The great strength of the book is in its worked examples, which have been chosen with a deep and realistic understanding of industrial processes. They perceptively demonstrate how very simple models can be used to great advantage to clarify the essential features of quite complex phenomena.

Work on the translation was relaxing for me because I did not find it desirable to suggest any changes in the original. The notation in the book did not appear to me to be sufficiently intuitive for an American or English-speaking student of engineering. This induced me to introduce certain changes in notation. In particular, I used the symbols k and μ for thermal conductivity and viscosity, respectively, in place of the original λ and η. Further, the symbols T, h, and U denote temperature (or temperature difference), film coefficient of heat transfer, and overall coefficient of heat transfer, in place of the original notations ϑ, α, and k.

It appears to me that the book should prove useful as an aid to courses in heat transfer at the advanced undergraduate and beginning graduate level at most American universities. This is due to the fact that it compactly covers the standard material and goes somewhat beyond it, particularly in Chapters 7-11, which discuss thermal explosions, distributed heat sources, moving heat sources, and nonsteady three-dimensional conduction processes. Particularly welcome is the clear (and correct!) account of the Stefan-Neumann problem of conduction in the presence of phase transition.

I should like to record here my thanks for the great help I received from the authors, Professor Grigull and Dr. Sandner, who carefully read the first typescript as well as the proofs, and make sure that my rendering of their text corresponded to their intentions. I also thank my secretary, Mrs. Jeanne Pion, who so efficiently typed the manuscript and succeeded in faultlessly transcribing the rather complex equations.

Last, but not least, I wish to express my thanks to Mr. W. Begell of Hemisphere Publishing Corporation who persuaded me to undertake this task and who made it possible for my wife, Alicja, to help me with the final editing of this text for the printer. Her help is also gratefully acknowledged.

J. Kestin

PREFACE TO THE ENGLISH-LANGUAGE EDITION

We gladly welcomed Professor Kestin's intention to translate our book into English, not least because we could hardly think of a more competent translator. Since our book uses exclusively SI units, it was not necessary to perform conversions because SI has now become widely accepted in the Anglo-Saxon literature. We have readily agreed to Professor Kestin's suggestions concerning changes in notation in order more closely to conform to that used in American books.

We hope that this English edition of our book will meet with a favorable reception. May it contribute to a better understanding of the laws of conduction. We thank Professor Kestin not only for his work on the translation but also for his cooperation.

U. Grigull
H. Sandner

INTRODUCTION TO THE
International Series in Heat and Mass Transfer

Springer-Verlag and Hemisphere Publishing Corporation have jointly started the *International Series in Heat and Mass Transfer*, which is affiliated with Springer-Verlag's new German-language series, *Wärme- und Stoffübertragung*, edited by Professor U. Grigull. *Heat Conduction* and *Wärmeleitung* are, respectively, the first volumes published in these two series.

The general goal of the series is to provide more detailed and updated coverage of topics presented in the classic work by Gröber, Erk, and Grigull, *Wärmeübertragung*, which was translated by Jerzy Moszynski in 1961. A comprehensive coverage of theory and practice in all major subjects of heat and mass transfer is planned, including convective heat and mass transfer, boiling and condensation, radiation, optical methods, liquid metals, and mass transfer.

It is most appropriate that heat transfer developments in the Federal Republic of Germany be included in this international series since German educators and researchers have had a profound influence on heat transfer pedagogy and practice in the English-speaking technical community. Strong research institutes at many universities continue this tradition of gaining new knowledge and transmitting the accumulated understanding. English-language translations of many forthcoming volumes in the German series *Wärme- und Stoffübertragung*, along with original works on current topics in heat and mass transfer, will be published in the series.

In spite of the increasing number of international conferences and symposia that require manuscripts in English, much important German work remains untranslated. There are no translation journals. Also, the foreign language capability of researchers in English-speaking countries is declining, largely due to changes in doctoral requirements. Accordingly, the English translations of comprehensive new texts should be welcomed.

The *International Series in Heat and Mass Transfer* will include introductory level texts of a tutorial nature, as well as advanced texts and reference books for students, engineers, and scientists working in research and advanced development.

It is fitting that the *International Series in Heat and Mass Transfer* is initiated with the translation of an important German textbook during the 300th anniversary of German immigration in North America. I trust that *Heat Conduction* and forthcoming translated volumes will be eloquent reminders of the technical roots that were extended, starting fifty years ago, to strongly influence American heat transfer theory and practice.

A. E. Bergles

FREQUENTLY USED SYMBOLS

Symbol	Meaning	SI unit (as example)
a	Thermal diffusivity $\quad a = k/\rho c_p$	m^2/s
A	Area	m^2
b	Coefficient of heat penetration $\quad b = (k\rho c_p)^{1/2}$	$Ws^{1/2}/m\,K$
c_p, c_v	Specific heats	$J/kg\,K$
d, D	Diameter	m
D	Coefficient of diffusion	m^2/s
g	Acceleration due to gravity	m/s^2
h	Coefficient of heat transfer (film coefficient)	$W/m^2\,K$
h	Specific enthalpy	J/kg
H	Enthalpy	J
I	Electric current	A
k	Thermal conductivity	$W/m\,K$
l, L	Length	m
m, M	Mass	kg
p	Pressure	Pa
p	Modulus	1

Symbol	Meaning	SI unit (as example)
$\dot{q}$	Heat flux	W/m^2
Q	Quantity of heat	J
$\dot{Q}$	Heat rate	W
r, R	Radius	m
s	Laplace variable	s^{-1}
s	Distance	m
S	Shape factor	m
t	Time	s
T	Temperature, temperature difference	°C or K
u	Circumference	m
U	Overall coefficient of heat transfer	W/m^2 K
V	Volume	m^3
x, y, z	Cartesian coordinates	m
X	Reference length	m
δ	Eigenvalue (plate)	1
δ	Distance	m
$\epsilon_w, \epsilon_k, \epsilon_\infty$	Figures of merit for fins	1 = W/W
ξ, η, ζ	Reduced coordinates	1 = m/m
μ	Dynamic viscosity	kg/m s = Pa s
ρ	Density	kg/m^3
ρ	Reduced radius	1 = m/m
τ	Time	s
ϕ	Angle	1 = m/m
ψ	Stream function	K

Dimensionless quantities are listed in Appendix D.

ONE

INTRODUCTORY REMARKS

1.1 FUNDAMENTAL CONCEPTS

The term *heat conduction* denotes the transport of energy as a result of molecular interactions under the influence of a nonhomogeneous temperature distribution. At every point in a body, which we at first regard as homogeneous and isotropic, it is possible to identify an instantaneous heat transfer rate $\dot{Q}(x, y, z)$ (measured in W in the SI system). When we refer this heat transfer rate to an element of area at right angles to its direction, we obtain the heat flux $\dot{q}$ (measured in W/m² in the SI system). The heat flux $\dot{q}$ is related to the temperature field $T(x, y, z)$ through the equation

$$\dot{q} = -k \ \text{grad} \ T \tag{1.1}$$

first introduced into science by Biot (1804, 1816) and Fourier (1822). Very precise measurements have confirmed the validity of this equation (within the constraints of its assumptions) without exceptions. In words, Eq. (1.1) states that the heat flux $\dot{q}$ is proportional to the gradient of temperature and is counterdirectional to it. The proportionality factor k in Eq. (1.1) is known as the thermal conductivity (W/m K in SI).

1.2 FOURIER'S EQUATION

We consider a lamina of thickness dx cut out at right angles to an arbitrarily chosen direction x inside a body of very large dimensions (Fig.

1

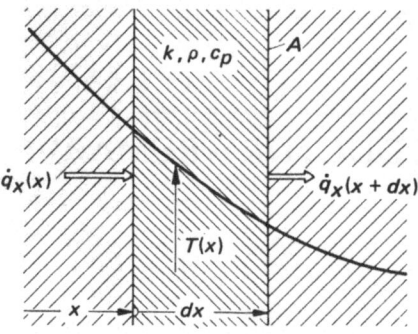

Figure 1.1 Derivation of Fourier's equation.

1.1). We assume that the heat fluxes that cross its two faces under the action of a temperature distribution $T(x)$ are $\dot{q}_x(x)$ and $\dot{q}_x(x+dx)$. The difference between the quantities of heat that cross the control surface A in time dt can be written

$$[\dot{q}_x(x) - \dot{q}_x(x+dx)]A\ dt = -\frac{d\dot{q}_x}{dx} A\ dx\ dt$$

where the series expansion

$$\dot{q}_x(x+dx) = \dot{q}_x(x) + \frac{d\dot{q}_x}{dx}\ dx$$

has been employed. In the absence of heat sources or heat sinks, this difference in heat flows serves to effect the enthalpy change $A\rho\ dh\ dx$ of the mass $A\rho\ dx$ enclosed inside the control volume $A\ dx$, where ρ denotes the density of the body under consideration. Noting that $dh = c_p\ dT$ and employing the Biot-Fourier phenomenological assumption $\dot{q}_x = -k\ dT/dx$ from Eq. (1.1), we obtain

$$-\frac{\partial \dot{q}_x}{\partial x} = \frac{\partial}{\partial x}\left(k\frac{\partial T}{\partial x}\right) = \rho c_p \frac{\partial T}{\partial t}$$

Since the body is homogeneous and isotropic, as assumed in Sec. 1.1, we conclude that the thermal conductivity k is independent of x or of any other space coordinate. If k is, further, assumed to be independent of temperature, we may write

$$\frac{\partial T}{\partial t} = a\frac{\partial^2 T}{\partial x^2} \tag{1.2}$$

with $a = k/(\rho c_p)$. Equation (1.2) is a differential equation that describes the nonsteady process of one-dimensional heat conduction. This—as well as the various modifications of this equation—is called the Fourier equation in honor of its discoverer. The material property $a = k/(\rho c_p)$ is called thermal diffusivity (SI unit: m^2/s). Equation (1.2) states that the local rate of temperature change $\partial T/\partial t$ is proportional to the curvature of the temperature distribution $T(x)$ which depends, intrinsically, on its second derivative $\partial^2 T/\partial x^2$. Spatially steep instantaneous temperature distributions $T(x)$ are the fastest to smooth out in time. The thermal diffusivity a provides us with a measure of the rate of this smoothing.

In the three-dimensional case, the enthalpy of the control volume in time dt is also affected by the differences in the heat flows in the y and z directions. Hence we may write

$$\rho c_p \frac{\partial T}{\partial t} = \frac{\partial}{\partial x}\left(k\frac{\partial T}{\partial x}\right) + \frac{\partial}{\partial y}\left(k\frac{\partial T}{\partial y}\right) + \frac{\partial}{\partial z}\left(k\frac{\partial T}{\partial z}\right) = \text{div}\ (k\ \text{grad}\ T)$$

(1.3)

If the thermal conductivity is independent of position and time, we may simplify Eq. (1.3) to the form

$$\frac{\partial T}{\partial t} = a\left(\frac{\partial^2 T}{\partial x^2} + \frac{\partial^2 T}{\partial y^2} + \frac{\partial^2 T}{\partial z^2}\right) = a\ \nabla^2 T \qquad (1.4)$$

where ∇^2 denotes the Laplacian operator.

If the control volume contains volumetric heat sources $\dot{q}'$ (SI unit: W/m^3) (when heat sinks are present, $\dot{q}'$ is negative), it becomes necessary to add this contribution to the right-hand side of Eq. (1.3). In the case of a temperature-independent thermal conductivity, we derive from Eq. (1.4) the relation

$$\frac{\partial T}{\partial t} = a\ \nabla^2 T + \frac{\dot{q}'}{\rho c_p} \qquad (1.5)$$

The distributed source $\dot{q}'(x, y, z, t)$ can be created by electric heating or by chemical, biological, as well as nuclear reactions.

1.3 INITIAL AND BOUNDARY CONDITIONS

The differential equation (1.5) with heat sources, or Eq. (1.4) without them, allows us to calculate the temperature distribution $T(x, y, z, t)$ if certain conditions are prescribed on the boundaries of the four-dimensional time-space continuum. These are the so-called initial and boundary conditions which must be supplemented with a specification of the time and space distribution of $\dot{q}'$, when Eq. (1.5) is relevant.

In many cases, the temperature distribution $T(x, y, z)$ is specified at

$t = 0$. In the simplest case this may be a constant $T = T_c$. As far as spatial distributions are concerned, we distinguish three cases:

1. Boundary Condition of the First Kind: The temperature $T_w = T_w(t)$ is prescribed at the boundaries of the body under consideration. We shall refer to them as the temperature variation "at the wall," subscript w. For example, these may occur in the form of a constant, $T_w = \text{const}$, or as a periodic function, $T_w = T_o \cos \omega t$, where ω is the circular frequency. In this case, we may wish to calculate the variation of the heat flux $\dot{q}_w = \dot{q}_w(t)$ at the wall.
2. Boundary Conditions of the Second Kind: In this case the heat flux at the wall, $\dot{q}_w = \dot{q}_w(t)$, is prescribed. For example, such a heat flux may be created by electric heating. Since according to Eq. (1.1), we must have

$$\dot{q}_w = -k \left(\frac{\partial T}{\partial n} \right)_w \tag{1.6}$$

the indication of the heat flux implies that the temperature gradient at the wall and normal to it is also prescribed. Here n denotes the normal to the wall and its sign is positive when pointing inward. The solution consists of finding the temperature distribution $T_w = T_w(t)$. If the wall is adiabatically insulated, i.e., impermeable to heat, we must have $\dot{q}_w = 0$, that is also $(\partial T/\partial n)_w = 0$. In such cases, the curve $T(n)$ merges with the normal at the wall.
3. Boundary Conditions of the Third Kind: Often, the solid body exchanges heat with a fluid medium surrounding it. This process of heat transfer is described by the assumption that

$$\dot{q}_w = h(T_\infty - T_w) \tag{1.7}$$

which was first employed by Sir Isaac Newton[†] in his paper on thermometry (1701). In this equation, T_∞ denotes the temperature of the surrounding fluid outside the layer adjacent to the wall, the so-called thermal boundary layer. The coefficient of proportionality h is known as the coefficient of heat transfer, or film coefficient (SI unit: W/m^2 K). Substituting Eq. (1.6) into (1.7), we derive the condition that

$$-\left(\frac{\partial T}{\partial n} \right)_w = \frac{T_\infty - T_w}{k/h} \tag{1.8}$$

whose geometric interpretation is given in Fig. 1.2. At randomly chosen

[†]Grigull, U., Newton's Law of Cooling. Abhandl. der Braunschweigischen Wiss. Ges., vol. 29, Braunschweig, 1978, pp. 7–31. Commentary on a paper by Newton written in 1701 (in German).

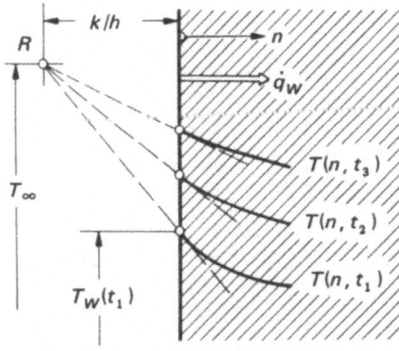

Figure 1.2 Geometric interpretation of the boundary condition of the third kind.

instants of time t_1, t_2, t_3, the tangents to the isotherms $T(n, t_i)$ measured at the wall pass through the fixed point R which is located at a distance k/h from the wall. It is possible to convert the boundary condition of the third kind, at least in steady-state problems, to one of the first kind by adding a hypothetical layer of thickness k/h to the solid body and assuming that the extended contour has a constant temperature ("method of auxiliary layer"). The coefficient of heat transfer h is not a thermophysical property of the fluid, unlike k, but depends in a complex way on the flow conditions in and the state of the surrounding fluid.

TWO

TRANSPORT PROPERTIES

The thermal conductivity k and the thermal diffusivity $a = k/(\rho c_p)$ were introduced in Secs. 1.1 and 1.2 as empirical factors and no mention was made of their order of magnitude. Nevertheless, our knowledge of the structure of matter allows us to provide at least qualitative indications.

2.1 THERMAL CONDUCTIVITY OF METALS AND ALLOYS

The electron theory of metals [2.1] proves that heat is conducted in crystalline metallic solids by the operation of two mechanisms. These are the lattice vibrations, whose energy quanta are called phonons, and the movement of conduction electrons. The two mechanisms operate in parallel and nearly independently so that the two contributions, the electronic contribution k_e and the lattice contribution k_l, may simply be added to obtain

$$k = k_e + k_l \tag{2.1}$$

As a rule, in metals, the term k_e dominates. The thermal conductivity is limited by the processes of scattering of both phonons and electrons. These occur through phonon-phonon and electron-phonon interactions. Scattering occurs also at the crystal boundaries, at crystal faults (e.g., dislocations) and foreign atoms such as impurities or components of alloys. If we imagine that the scattering processes are connected in series, we may sum up the resistivities $\bar{\rho} = 1/k$, that is, the reciprocals of the conductivities. Specifically, the electronic resistivity $\bar{\rho}_e = 1/k_e$ (Matthiesen's rule) can be thought of as consisting of two contributions, that due to the scattering by phonons ($\bar{\rho}_p$) and by impurities, grain boundaries, and generally lattice faults of all kinds ($\bar{\rho}_i$, subscript i for "impurus" or "impurity"). Hence thermal electronic resistivity is given by

$$\bar{\rho}_e = \frac{1}{k_e} = \bar{\rho}_p + \bar{\rho}_i \tag{2.2}$$

The electron theory of metals shows that $\bar{\rho}_p = \alpha T^2$ and $\bar{\rho}_i = \beta/T$ at low temperatures. Here α and β are two material constants, the constant β characterizing the number and nature of faults. Thus, the thermal conductivity of a metal at low temperature can be described by the equation

$$k \approx k_e = \frac{1}{\bar{\rho}_p + \bar{\rho}_i} = \frac{1}{\alpha T^2 + \beta/T} \tag{2.3}$$

The graph in Fig. 2.1 represents a schematized diagram of the thermal conductivity, k, of copper. At low temperatures, the term β/T in Eq. (2.3)

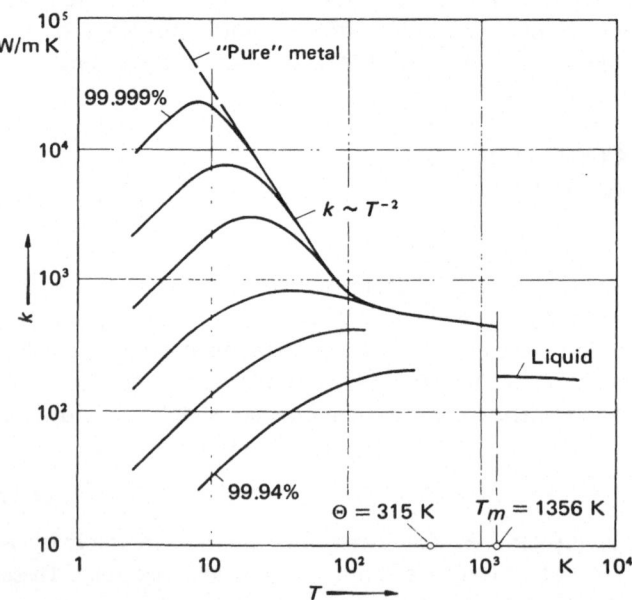

Figure 2.1 Thermal conductivity k of copper in terms of the absolute temperature T. The melting point is denoted by T_m and the characteristic temperature is Θ. The degree of purity is indicated in percentages. The diagram is schematized.

dominates, imparting a practically linear relationship to $k(T)$. This linear behavior persists to progressively higher temperatures as the number of scattering centers increases (large values of β). At higher temperatures the term αT^2 gains the upper hand, especially in the case of very pure metals (very small value of β). Thus, the maximum implied in Eq. (2.3) shifts in the direction of lower temperatures. In "pure" metals ($\beta \to 0$) the thermal conductivity k behaves as T^{-2}, as indicated in Fig. 2.1. In any case, even a very small number of scattering centers induces a drop in k for sufficiently low values of T.

Differentiating Eq. (2.3) with respect to T, we can establish the location T_{max} of the maximum value k_{max}. Thus,

$$T_{max} = \left(\frac{\beta}{2\alpha}\right)^{1/3} \tag{2.4}$$

We insert this value into Eq. (2.3) and introduce the dimensionless quantities $k^* = k/k_{max}$ and $T^* = T/T_{max}$, and thus obtain the universal relation

$$k^* = \frac{3}{T^{*2} + 2/T^*} \tag{2.5}$$

whose good performance in the range $0 < T^* < 1.5\, T_{max}$ has been verified by comparison with experiment [2.2]. The reference values T_{max} and k_{max} are not known from first principles, but can be obtained on the basis of suitably performed measurements.

The preceding considerations are valid for temperature $T \ll \Theta$, where Θ denotes a characteristic temperature. In Fig. 2.1 this characteristic temperature has been given its value $\Theta = 315$ K for copper. At higher temperatures ($T \approx \Theta$), it is possible to recommend two empirical equations, Eq. (2.2):

for $\qquad 0.3\Theta < T < 0.86\Theta \qquad k = 0.989 k_\Theta \exp \dfrac{0.0117}{(T/\Theta)^{2.5}} \tag{2.6}$

for $\qquad 0.86\Theta < T < 3\Theta \qquad k = k_\Theta \left(1.05 - 0.05\, \dfrac{T}{\Theta}\right) \tag{2.7}$

In these equations k_Θ denotes the thermal conductivity at the characteristic temperature Θ; its values for some metals are listed in Table 2.1. The column denoted by $k_{\Theta i}$ in the same table contains values of the thermal conductivity of metals containing 0.5% impurities, whereas k_Θ refers to the pure metals. Values for intermediate degrees of impurity can be obtained by linear interpolation. Table 2.1 lists, further, the atomic number Z, the relative atomic weight A_r, and the melting point T_m. The characteristic temperatures listed here have been obtained by curve-fitting to measure-

Table 2.1 Atomic number Z, relative atomic weight A_r, characteristic temperature Θ, melting point T_m, thermal conductivity k_Θ and $k_{\Theta i}$ for metals; Eqs. (2.6) and (2.7). Reference [2.2].

Metal	Z	A_r	Θ, K	T_m, K	k_Θ, W/m K	$k_{\Theta i}$, W/m K
Ag	47	107.870	215	1234	420	380
Al	13	26.982	390	933.2	230	210
Au	79	196.97	170	1336.2	348	320
Cd	48	112.40	220	594.2	113	84
Cu	29	63.54	315	1356	414	330
Ir	77	192.2	285	2716	160	130
Mg	12	24.31	290	923	170	140
Pb	82	207.19	88	600.58	47	35
Pd	46	106.4	275	1825	72	60
Rh	45	102.91	370	2233	160	130
Ti	22	47.90	350	1953	25	15
Tl	81	204.4	100	576.2	60	44
W	74	183.85	310	3653	170	160
Zn	30	65.37	250	692.7	140	105
Zr	40	91.22	280	2125	26	19
Co	27	58.93	385	1765	130	100
Cr	24	51.99	485	2118	84	65
K	19	39.10	100	336.8	120	100
Li	3	6.939	400	453.7	80	65
Mo	42	95.94	380	2883	150	120
Na	11	22.99	150	371.0	150	120
Pt	78	195.1	225	2042	80	65
Rb	37	85.47	85	312.04	75	60
Re	75	186.2	300	3453	55	45

ments. These are close to the Debye temperatures Θ_D obtained from measurements of specific heats.

The relationship between the electronic contribution of thermal conductivity to that of electric conductivity (which, generally speaking, is much easier to measure) is brought to evidence by the relation

$$\frac{k}{\sigma T} = L \tag{2.8}$$

derived by Wiedemann, Franz, and Lorenz. Here σ denotes electric conductivity (SI unit: A/V m), and L is the Lorenz constant. According to Sommerfeld, the theoretical value L_0 of Lorenz's constant is given as

$$L_0 = \frac{\pi^2}{3}\left(\frac{k}{e}\right)^2 = 2.45 \times 10^{-8} \; \frac{V^2}{K^2} \tag{2.9}$$

Here k is Boltzmann's constant, and e is the electron charge. In a somewhat extended form, this equation can be successfully applied to alloy steels.

Example 2.1 According to Eq. (2.7) and Table 2.1 thermal conductivity k_{20} of pure silver at $20°C \approx 293$ K is $k_{20} = 420[1.05 - 0.05(293/215)]$ W/m K $= 412$ W/m K. In the presence of 0.5% impurities, we would obtain $k_{20} = 372$ W/m K.

2.2 TRANSPORT PROCESSES IN DILUTE GASES

For simplicity we consider a monatomic gas of moderate density. In such circumstances molecules interact only during a collision. We denote the number of particles in a volume element by N_V and their mean velocity by $\bar{w}$. We may assume by way of approximation that the number of particles which move at any instant in one of the three coordinate directions is equal to $(\frac{1}{3})N_V$. Thus, the flux of particles (number of particles per unit area and time) moving in this direction becomes $(\frac{1}{3})N_V\bar{w}$. Now let G denote some property of these particles (such as their energy or momentum) whose distribution in the given direction is described by the function $G(y)$ and let $\bar{l}$ denote the mean free path, that is, an average distance traveled by a particle without changing its value of G. Then, the flux of property G in the given direction can be expressed as

$$\Gamma = -\frac{\bar{w}lN_V}{3}\frac{dG}{dy} \qquad (2.10)$$

In a perfect gas $N_V = p/(kT)$, where k is Boltzmann's constant. The kinetic theory of gases leads to the expression

$$\bar{w} = \left(\frac{8kT}{\pi m}\right)^{1/2} \qquad (2.11)$$

for the mean velocity, and to

$$\bar{l} = \frac{1}{\sqrt{2}\pi\sigma^2 N_V} \qquad (2.12)$$

for the mean free path. Here m denotes the mass of a particle and σ is the effective cross section for a collision.

To describe the transport of energy by conduction, we put $dG = du = mc_v\, dT$, where u denotes the mean energy of a particle and c_v is its isochoric specific heat. According to Eq. (2.10), Γ represents the heat flux $\dot{q}_w$, so that by comparing this expression with the linear assumption of

Biot-Fourier (here $\dot{q} = \dot{q}_y = -k dT/dy$) we can derive an expression for the thermal conductivity k in the form

$$k = \frac{\rho \bar{w} \bar{l} c_v}{3} = \frac{1}{\sigma^2} \left(\frac{k^3 T}{\pi^3 m} \right)^{1/2} \tag{2.13}$$

Here $\rho = m N_V$ is the density. Furthermore, for a monatomic gas which possesses only translational degrees of freedom, we may put $c_v = \frac{3}{2}R$, where R is the gas constant ($mR = k$). According to Eq. (2.13), the thermal conductivity k is independent of pressure and proportional to $T^{0.5}$.

According to Eq. (2.13), the thermal diffusivity $a = k/(\rho c_p)$ behaves as

$$a \sim \frac{T^{1.5}}{p} \tag{2.14}$$

It is possible to perform an analogous argument for the transport of momentum, which results in viscosity, and for the transport of mass, which gives rise to diffusion. When this is done, it is possible to obtain expressions for the dynamic viscosity μ (SI unit: $Ns/m^2 = Pa\,s$), the kinematic viscosity $\nu = \mu/\rho$ (SI unit: m^2/s) and the coefficient of diffusion D (SI units m^2/s). The latter actually represents the coefficient of self-diffusion in this simplified line of reasoning. Thus we would be led to the estimates

$$\mu \sim T^{0.5} \qquad \nu \sim \frac{T^{1.5}}{p} \qquad D \sim \frac{T^{1.5}}{p} \tag{2.15}$$

The pressure dependence that occurs in the preceding estimates is confirmed satisfactorily by measurements in the range of low pressures (below 10 bar). However, the measured rate of change with temperature is different. Even for monatomic gases the exponents are larger and closer to

$$k \sim T^{0.7} \qquad \mu \sim T^{0.7} \qquad D \sim T^{1.7}$$

This is also confirmed by Table 2.2. The group $Pr = \mu c_p/k$ listed in the table, known as the Prandtl number, plays an important part in the transfer of heat by convection. According to prediction, the Prandtl number is only a very weak function of temperature and is independent of pressure.

2.3 HEAT CONDUCTION THROUGH LAMINATED BODIES

In isotropic bodies discussed so far (Chapter 1), the heat flux and the negative temperature gradient from Eq. (1.1) are collinear; the thermal conductivity k is a scalar. Generally speaking, in an anisotropic body the preceding two vectors enclose an angle and the thermal conductivity has

Table 2.2 Transport properties of gases at moderate pressures[a]

μ_0 = viscosity; k_0 = conductivity; D_0 = coefficient of self-diffusion at 101,325 Pa; $Pr_0 = \mu_0/c_{po}k_0$ is the Prandtl number. All values for T_0 = 273.15 K = 0°C. Between 0° and 1000°C we may put $Z(T) = Z_0(T/T_0)^n$, where Z_0 denotes the tabulated values and n the associated exponent, also tabulated.

Gas	μ_0, μPa s	n	k_0, mW/K m	n	D_0, cm^2/s	n	Pr_0	n
Ar	21.45	0.693	16.8	0.693	0.158	1.70	0.666	0
H_2	8.50	0.648	171	0.690	1.26	1.66	0.700	0.007
D_2	11.70	0.648	117	0.720	0.862	1.66	0.700	0.011
O_2	19.69	0.672	25.2	0.802	0.181	1.68	0.701	0.019
N_2	16.80	0.666	24.1	0.760	0.177	1.68	0.700	0.015
Air	17.42	0.669	24.6	0.759	0.178	1.68	0.701	0.013
CO	16.80	0.672	24.2	0.779	1.177	1.68	0.700	0.016
CO_2	14.40	0.774	16.7	1.04	0.0962	1.78	0.725	0.027
NO	18.32	0.674	25.0	0.784	0.180	1.69	0.702	0.016
N_2O	14.33	0.794	17.1	1.06	0.0958	1.80	0.727	0.026
Cl_2	13.17	0.81	8.88	0.877	0.0547	1.81	0.717	0.009
SO_2	13.79	0.806	11.3	1.02	0.0611	1.81	0.730	0.021
CH_4	10.62	0.716	30.6	1.256	0.194	1.73	0.720	0.047
C_2H_6	8.96	0.788	20.4	1.423	0.0877	1.79	0.751	0.034
C_3H_8	7.90	0.807	17.2	1.45	0.0528	1.81	0.769	0.025
n-C_4H_{10}	7.20	0.908	15.7	1.53	0.0368	1.90	0.780	0.018
n-C_5H_{12}	6.63	0.872	14.4	1.49	0.0272	1.87	0.786	0.015
n-C_6H_{14}	6.73	0.910	12.5	1.52	0.0200	1.90	0.791	0.012
C_6H_6	7.21	0.922	10.3	1.58	0.0274	1.92	0.775	0.022
c-C_6H_{12}	6.65	0.858	11.2	1.65	0.0234	1.86	0.784	0.018
C_2H_2	10.01	0.75	21.9	1.38	0.113	1.76	0.739	0.039
C_2H_4	9.92	0.767	20.7	1.34	0.104	1.77	0.739	0.038
CH_3OH	9.19	0.948	17.2	1.47	0.0853	1.94	0.741	0.035
C_2H_5OH	8.29	0.898	16.5	1.47	0.0534	1.89	0.766	0.024

Numerical example: The thermal conductivity of air at 100°C ≈ 373.15 K is equal to k_{100} = 24.6 × (373.15/273.15)$^{0.759}$ mW/m K = 31.2 mW/m K = 0.0312 W/m K.

[a]Müller, R., *Chem. Ing. Tech.*, vol. 40, pp. 344–349, 1968.

the mathematical properties of a tensor. Above all, anisotropy is observed in crystals; it occurs also in naturally or artificially laminated bodies such as wood, minerals, plywood, plates used in the construction industry, laminated sheets, and so forth. In what follows, we shall restrict ourselves to the discussion of laminated structures and calculate the heat fluxes normal and at right angles to the laminae. Only steady-state conduction is discussed.

2.3.1 Heat Flow Normal to Lamina

We consider the body depicted in Fig. 2.2;. it consists of m individual laminae whose thicknesses are δ_1, δ_2, ..., δ_m and whose corresponding thermal conductivities are k_1, k_2, ..., k_m. We suppose that the two external planes are kept at temperatures T_I and T_{II} ($T_I > T_{II}$). In such circumstances there will arise a definite heat flux, equal through every layer, and we may write that

$$\dot{q} = \frac{k_1}{\delta_1} (T_I - T_a) = \frac{k_2}{\delta_2} (T_a - T_b) = \cdots = \frac{k_n}{\delta} (T_I - T_{II})$$

Here, T_a, T_b, ..., are the temperatures at the internal planes of contact and

$$\delta = \sum_{j=1}^{m} \delta_j$$

is the total thickness. The effective thermal conductivity normal to the layers is denoted by k_n. By the elimination of the intermediate temperatures T_a, T_b, ..., etc., we establish the relation

$$k_n = \frac{\delta_1 + \delta_2 + \cdots}{\delta_1/k_1 + \delta_2/k_2 + \cdots} = \frac{\delta}{\sum\limits_{j=1}^{m} \delta_j/k_j} \qquad (2.16)$$

This equation represents resistances connected in series and consists of a sum of reciprocal conductivities. To display this explicitly, Eq. (2.16) can also be written

$$\frac{\delta}{k_n} = \frac{\delta_1}{k_1} + \frac{\delta_2}{k_2} + \cdots = \sum_{j=1}^{m} \frac{\delta_j}{k_j} \qquad (2.17)$$

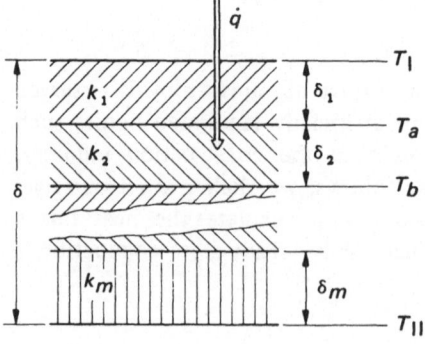

Figure 2.2 Heat flux normal to the layers.

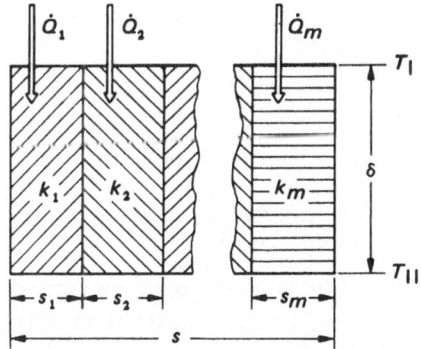

Figure 2.3 Heat flux parallel to the layers.

2.3.2 Heat Flow Parallel to the Layers

We consider the structure, shown in Fig. 2.3, which consists of m layers of thickness, s_1, s_2, ..., s_m of materials whose respective thermal conductivities are k_1, k_2, ..., k_m. The total thickness is $s = \Sigma_{j=1}^{m} s_j$.

When two parallel planes normal to the layers and separated by a distance δ are kept at constant temperatures T_I and T_{II} ($T_I > T_{II}$), we can represent the partial heat transfer rates in the form

$$\dot{Q}_1 = k_1 \frac{s_1 l}{\delta} (T_I - T_{II}), \quad \dot{Q}_2 = k_2 \frac{s_2 l}{\delta} (T_I - T_{II}), \quad \text{etc.} \qquad (2.18)$$

Here l denotes a constant length normal to the plane of the drawing. The partial heat flows add up to a total heat transfer rate

$$\dot{Q} = \dot{Q}_1 + \dot{Q}_2 + \cdots = k_p \frac{s l}{\delta} (T_I - T_{II}) \qquad (2.19)$$

where k_p is the effective thermal conductivity parallel to the layers. We substitute the expressions from Eq. (2.18)–(2.19) and obtain

$$\dot{Q} = \frac{1}{\delta} (T_I - T_{II})(k_1 s_1 + k_2 s_2 + \cdots)$$

Upon comparing with the right-hand side of Eq. (2.19) we get

$$k_p = \frac{k_1 s_1 + k_2 s_2 + \cdots}{s_1 + s_2 + \cdots} = \frac{\sum_{j=1}^{m} k_j s_j}{s} \qquad (2.20)$$

2.3.3 Comparison between k_n and k_p

The ratio of the two effective thermal conductivities is given by

$$\frac{k_p}{k_n} = \frac{\sum\limits_{j=1}^{m} k_j s_j \sum\limits_{j=1}^{m} \delta_j/k_j}{s\delta}$$

In the same structure s becomes identical with δ, and we obtain

$$\frac{k_p}{k_n} = \frac{1}{\delta^2}(k_1\delta_1 + k_2\delta_2 + k_2\delta_2 + \cdots)\left(\frac{\delta_1}{k_1} + \frac{\delta_2}{k_2} + \cdots\right) \geqslant 1 \quad (2.21)$$

The ratio k_p/k_n becomes equal to unity only in the special case when $k_1 = k_2 = \cdots$ (isotropic body). In all other cases this ratio is greater than unity as is easy to verify by performing the multiplication indicated in Eq. (2.21).

In practical applications we often encounter laminated materials which consist of m layers of two different components, each of thickness δ_1 or δ_2 and thermal conductivity k_1 or k_2. The total thickness is then $\delta = m\delta_1 + m\delta_2 = m\delta_1(1 + a)$, where $a = \delta_2/\delta_1$ is the constant thickness ratio. In such cases we obtain

$$\frac{k_p}{k_n} = \frac{k_1 + ak_2}{(1 + a)^2}\left(\frac{1}{k_1} + \frac{a}{k_2}\right) = 1 + \frac{a(k^* - 2)}{(1 + a)^2} \quad (2.22)$$

The contraction k^* has here been defined as $k^* = (k_1/k_2) + (k_2/k_1) \geqslant 2$. Since k^* is symmetric with respect to the indices 1 and 2, it follows that interchanging the two different materials changes nothing in the properties of the compound. The maximum value implied in Eq. (2.22) occurs when $a = 1$ or $\delta_1 = \delta_2$ and is

$$\left(\frac{k_p}{k_n}\right)_{max} = \frac{k^* + 2}{4} \quad (2.23)$$

Example 2.2 We consider a laminated material made up of two unknown components, but consisting of layers of equal thickness ($a = 1$). Measurements yielded the value $k_p/k_n = 2$. It is seen from Eq. (2.23) that we then must have $k^* = 6$ and $k_1/k_2 = 5.83$ or 0.17.

Example 2.3 A sandwich plate consists of copper plates ($k_1 = 390$ W/m K) separated by layers of foam rubber of equal thickness ($k_2 = 0.03$ W/m K). We calculate that $k_1/k_2 = 13,000$ so that $k^* \approx 13,000$. Hence $k_p/k_n \approx 3,250$.

2.4 ORDER OF MAGNITUDE OF HEAT FLUXES AND COEFFICIENTS

In preliminary design calculations it is often advantageous to estimate the correct order of magnitude without much preparation. For this reason, we

Table 2.3 Heat fluxes $\dot{q}$ in W/m^2 or W/cm^2

Terrestrial heat flux	$0.063 \ W/m^2$
Barely perceptible heat radiation from human skin	$40 \ W/m^2$
Threshold of pain of thermal radiation	$1500-2500 \ W/m^2$
Heat loss from human body	$50 \ W/m^2$
General radiation from cloudless atmosphere	$200 \ W/m^2$
Electric heating of highways in winter (Federal Republic of Germany)	$70-350 \ W/m^2$
Radiant heat from ceiling	$100 \ W/m^2$
Heating of water (at the heating element)	$500-800 \ W/m^2$
Sun in middle of summer	$500-800 \ W/m^2$
Solar constant	$1326 \ W/m^2$
Heating of containers, domestic appliances	$1-8 \ W/cm^2$
Supercritical boilers, high-output heat pipe	$50 \ W/cm^2$
Fuel element in nuclear reactor	$100 \ W/cm^2$
Cooling of rocket nozzles	$4500 \ W/cm^2$

Table 2.4 Thermal conductivities, k, and thermal diffusivities, a

Material	k, W/m K	a, $10^{-6} \ m^2/s$
Metals	5 –400	3 –100
Inorganic solids	0.5 –10	0.5–1
Rocks	1.6 –2.9	1 –1.4
Organic solids	0.1 –1	0.1
Liquids	0.1 –1	0.1
Gases	0.01–0.2	3 –100

Table 2.5 Heat transfer coefficients (film coefficients) h

	h, W/m^2 K
Natural convection	
Gases	3–20
Water	100–600
Boiling water	1000–20,000
Forced convection	
Gases	10–100
Viscous liquids	50–500
Water	500–10,000
Condensing steam	1000–100,000

Table 2.6 Coefficients of heat penetration
$b = k/a^{1/2} = (k\rho c_p)^{1/2}$ in W s$^{1/2}$/m^2 K

Copper	36,000	Sand	1,200
Iron	15,000	Wood	400
Concrete	1,600	Foam material	40
Water	1,400	Gases	6

have indicated in Tables 2.3–2.6 a series of representative values. In addition to the quantities $\dot{q}$, k, a, and h already utilized, we have listed the so-called coefficient of heat penetration defined as

$$b = \frac{k}{a^{1/2}} = (k\rho c_p)^{1/2}$$

(SI units: W s$^{1/2}$/m^2 K). This coefficient measures, for example, the quantity of heat which penetrates into a body during a given time interval when the temperature of its surface has been suddenly raised by a given amount.

It is worth noting in Table 2.4 that the thermal diffusivities a of metals and gases are of the same order of magnitude. It follows that temperature differences decay equally fast in a layer of metal as they do in a layer of air.

THREE

ONE-DIMENSIONAL, STEADY-STATE HEAT CONDUCTION

3.1 INTRODUCTORY REMARKS

We consider an isotropic body not containing heat sources and whose temperature distribution $T(x, y, z)$ does not depend on time ($\partial T/\partial t = 0$; steady state). According to Eq. (1.3), we then have

$$\text{div} \, (k \, \text{grad} \, T) = 0 \tag{3.1}$$

If the thermal conductivity is independent of temperature, this equation simplifies to

$$\nabla^2 T = 0 \tag{3.2}$$

which is Laplace's potential equation. Depending on the form of the Laplace operator, we can write this equation in the compact form

$$\frac{d^2 T}{dr^2} + \frac{n}{r} \frac{dT}{dr} = 0 \tag{3.3}$$

where r denotes the single space coordinate (referred to an arbitrary point in a plate; for a cylinder or sphere, r denotes the radius), and $n = 0, 1$, or 2 for plates, cylinders, and spheres, respectively. We assume that the boundary conditions are prescribed in the form of two temperatures (boundary conditions of the first kind), namely

$$T = T_i \qquad \text{for } r = r_i \text{ (inner)}$$

$$T = T_e \qquad \text{for } r = r_e \text{ (outer)}$$

We denote the thickness of the plate by $\delta = r_e - r_i$ and write down the three solutions of Eq. (3.1) which have been sketched in Fig. 3.1. These are

$$\frac{T - T_e}{T_i - T_e} = \frac{r_e - r}{r_e - r_i} = \frac{r_e - r}{\delta} \qquad \text{(plate)} \qquad (3.4)$$

$$\frac{T - T_e}{T_i - T_e} = \frac{\ln (r_e/r)}{\ln (r_e/r_i)} \qquad \text{(cylinder)} \qquad (3.5)$$

$$\frac{T - T_e}{T_1 - T_e} = \frac{1/r - 1/r_e}{1/r_i - 1/r_e} \qquad \text{(sphere)} \qquad (3.6)$$

The equations for the total heat transfer rate $\dot{Q}$ from r_i to r_e assume the forms:

$$\dot{Q} = kA \frac{T_i - T_e}{r_e - r_i} = k \frac{A}{\delta} (T_i - T_e) \qquad \text{(plate)} \qquad (3.7)$$

$$\dot{Q} = \frac{2\pi k(T_i - T_e)}{\ln (r_e/r_i)} \qquad \text{(cylinder)} \qquad (3.8)$$

$$\dot{Q} = \frac{4\pi k(T_i - T_e)}{1/r_i - 1/r_e} \qquad \text{(sphere)} \qquad (3.9)$$

Here A denotes the area of the flat plate, and L is the cylinder length. Equations (3.8) and (3.9) allow us to recognize that the cases of thin hollow cylinders or spheres can be treated by the same equations as those of the flat plate. This can be seen from the expansions:

$$\ln \frac{\bar{r} + (\delta/2)}{\bar{r} - (\delta/2)} = \ln \left(1 + \frac{1}{2}\frac{\delta}{\bar{r}}\right) - \ln \left(1 - \frac{1}{2}\frac{\delta}{\bar{r}}\right) = \frac{\delta}{\bar{r}} + O\left(\frac{\delta^3}{\bar{r}^3}\right)$$

$$\frac{1}{[1/(\bar{r} - \frac{1}{2}\delta)] - [1/(\bar{r} + \frac{1}{2}\delta)]} = \frac{\bar{r}^2 - \frac{1}{4}\delta^2}{\delta} = \frac{\bar{r}^2}{\delta} - O\left(\frac{\delta^2}{\bar{r}^2}\right)$$

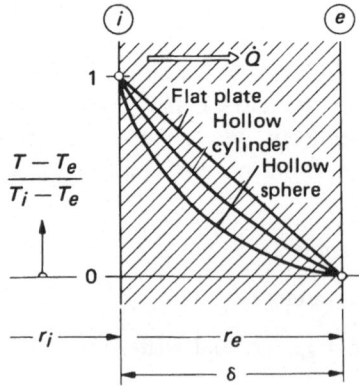

Figure 3.1 Steady-state temperature distribution across a flat plate, hollow cylinder, and hollow sphere.

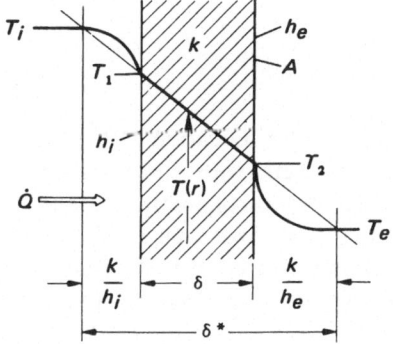

Figure 3.2 Derivation of Péclet's equation.

In the two cases, the area A in Eq. (3.7) is replaced by the area of the mean radius $\bar{r}$, that is, by

$$A = 2\pi\bar{r}L \quad \text{(cylinder)}$$

or

$$A = 4\pi\bar{r}^2 \quad \text{(sphere)}$$

If in a sphere $r_e \to \infty$ or $1/r_e \to 0$ ("infinite" body), we obtain

$$\dot{Q} = 4\pi k r_i (T_i - T_e) \tag{3.10}$$

where T_e is the temperature at a very large distance from the center.

3.2 PECLET'S EQUATIONS

We now propose to extend the equations derived in Sec. 3.1, Eqs. (3.7)–(3.10), to the case of boundary conditions of the third kind. The diagram in Fig. 3.2 depicts the process of heat transfer across a wall separating two fluids in which the temperatures T_i and T_e are maintained, as shown. The respective film coefficients are h_i and h_e. The temperatures at the wall differ from T_i and T_e, and are denoted by T_1 and T_2, respectively. The total, steady-state heat transfer rate $\dot{Q}$ through area A can be written

$$\dot{Q} = h_i A(T_i - T_1) = k\frac{A}{\delta}(T_1 - T_2) = h_e A(T_2 - T_e) \tag{3.11}$$

where Eq. (1.7) has been made use of. If we now solve each of the above three equations for the "driving" temperature difference and add all three, we are led to the result that

$$\dot{Q} = \frac{A(T_i - T_e)}{1/h_i + \delta/k + 1/h_e} = UA(T_i - T_e) \tag{3.12}$$

The quantity U, defined by

$$\frac{1}{U} = \frac{1}{h_i} + \frac{\delta}{k} + \frac{1}{h_e} \qquad \text{(plate)} \qquad (3.13)$$

is known as the overall coefficient of heat transfer. Equation (3.12) can be interpreted by stating that we are dealing here with three resistances connected in series so that the reciprocals of the conductivities must be added to one another. The overall coefficient of heat transfer U has the same dimension as that of h. If the wall were to consist of m layers of thickness δ_j and thermal conductivity k_j, we would replace Eq. (3.13) by the relation

$$\frac{1}{U} = \frac{1}{h_i} + \sum_{j=1}^{m} \frac{\delta_j}{k_j} + \frac{1}{h_e} \qquad \text{(plate)} \qquad (3.14)$$

According to Fig. 3.2, we may introduce a fictitious wall thickness

$$\delta^* = \frac{k}{h_i} + \delta + \frac{k}{h_e}$$

(cf. method of auxiliary wall mentioned in Sec. 1.3). Then the total heat transfer rate can be expressed in the form

$$\dot{Q} = \frac{Ak(T_i - T_e)}{\delta^*}$$

In the case of hollow cylinders and hollow spheres, the analogs of Eq. (3.12) assume the respective forms:

$$\dot{Q} = \frac{2\pi L(T_i - T_e)}{1/h_i r_i + \sum\limits_{j=1}^{m} \{[\ln (r_{2j}/r_{1j})]/k_j\} + 1/h_e r_e} \qquad \text{(cylinder)}$$

$$(3.15)$$

$$\dot{Q} = \frac{4\pi(T_i - T_e)}{1/h_i r_i^2 + \sum\limits_{j=1}^{m} [(1/k_j)(1/r_{1j} - 1/r_{2j})] + 1/h_e r_e^2} \qquad \text{(sphere)}$$

$$(3.16)$$

In Eqs. (3.15) and (3.16), r_i and r_e denote the radii of the innermost and outermost boundaries of the hollow cylinder and sphere, respectively, whereas r_{2j} and r_{1j} refer to the boundaries of each layer whose thermal conductivity is then denoted by k_j.

Equations (3.12), (3.15), and (3.16) are known as Péclet's equations in honor of their discoverer. The equations can be used to estimate the contribution made by each term in the denominator and so to decide if and where improvements can be made.

Example 3.1 The refractory insulation of an industrial boiler has a thickness $\delta = 0.25$ m and a thermal conductivity $k = 1$ W/m K. The effect of the steel wall can be neglected. Assuming $h_i = 100$ W/m^2 K and $h_e = 10$ W/m^2 K, we calculate from Eq. (3.14):

$$\frac{1}{h_i} = \frac{1}{100} \frac{m^2 \ K}{W} = 0.01 \frac{m^2 \ K}{W}$$

$$\frac{\delta}{k} = \frac{0.25 \ m}{1 \ W/m \ K} = 0.25 \frac{m^2 \ K}{W}$$

$$\frac{1}{h_e} = \frac{1}{10} \frac{m^2 \ K}{W} = 0.10 \frac{m^2 \ K}{W}$$

$$\frac{1}{U} \qquad = \qquad 0.36 \frac{m^2 \ K}{W}$$

If the inner temperature is $T_i = 700°C$ and the outer one is $T_e = 20°C$, we obtain the heat flux

$$\dot{q} = \frac{\dot{Q}}{A} = U(T_i - T_e) = \frac{(700 - 20)K}{0.36 \ m^2 \ K/W} = 1.89 \frac{kW}{m^2}$$

In order to reduce this heat loss, it is proposed to add an insulation of $\delta_1 = 0.05$ m and $k_1 = 0.1$ W/m K. In this manner the value of $1/U$ increases by $\delta_1/k_1 = 0.5$ m^2 K/W yielding $1/U = 0.86$ m^2 K/W and $\dot{q} = 791$ W/m^2.

Example 3.2 The intermediate temperatures of the wall of the preceding example can be calculated with the aid of Eq. (3.11). They can also be determined graphically in a simple manner, as shown in Fig. 3.3. We plot T against $1/h_i + \delta/k + 1/h_e$. The straight line a corresponds to the case without the insulating layer. The external temperature turns out to be 208°C. The heat flux is proportional to $\tan \beta$. The addition of the layer of insulation (straight line b) causes the external wall temperature to drop to 98°C. The temperature at the boundary between insulation and refractive layer is 495°C and the heat flux becomes proportional to $\tan \gamma$. The temperature drop at the inner wall caused by h_i is very small.

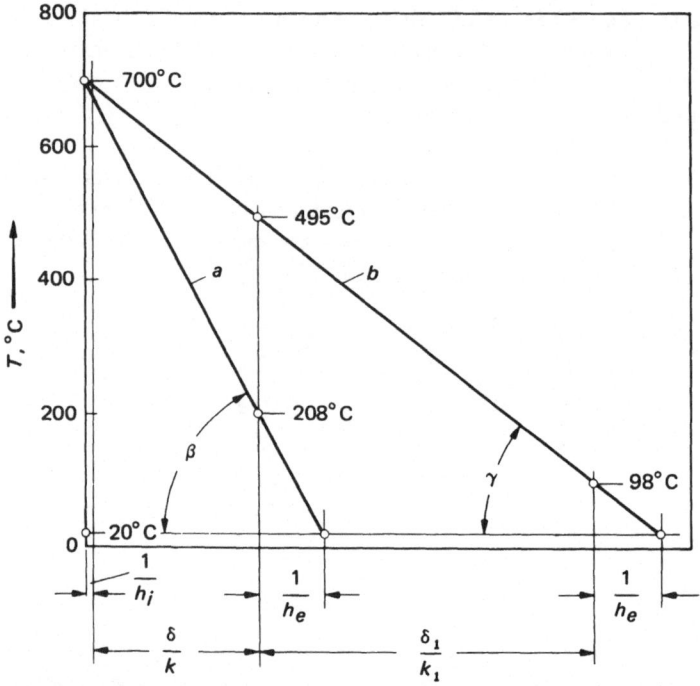

Figure 3.3 Graphical determination of the temperatures in the wall of a furnace with (straight line *a*) and without (straight line *b*) insulation.

A layer of thickness $\delta = r_e - r_i$ placed on a cylinder or sphere need not reduce the heat flux; under certain circumstances it may cause it to increase. This is due to the fact that the insulating layer increases the resistance to heat flow, but at the same time the area through which heat is lost increases. Neglecting the resistance of the inner film $(1/h_i = 0)$, we can write for the cylinder that

$$\dot{Q} = \frac{2\pi kL(T_i - T_e)}{\ln(1 + \delta/r_i) + k/[h_0 r_i(1 + \delta/r_i)]}$$

The derivative of $\dot{Q}$ with respect to δ shows that its value is positive in the range

$$\frac{h_e r_e}{k} < 1 \tag{3.17}$$

This signifies that an additional layer of insulation causes the heat loss $\dot{Q}$ to increase as long as $r_e < k/h_0$. In the case of the sphere this occurs for $r_e < 2k/h_e$. Examples of cases when we seek to increase $\dot{Q}$ by the addition of an insulation are provided by insulated, free-hanging electric conductors and pipes covered with frost in air-conditioning installations.

3.3 QUASI-STEADY HEAT CONDUCTION

In many cases, the temperature changes occur so slowly that it is permissible to assume that they are steady state at every instant (quasi-steady approximation). Thus, several important applications can be analyzed with the aid of the relations derived so far.

3.3.1 Heat Loss in Lagged Pipes

In the case of pipes it is almost always permissible to put $h_i \gg h_e$ in Eq. (3.15). This makes the inner wall temperature equal to that of the flowing fluid. The heat loss per unit pipe length L becomes

$$\dot{Q}_L = \frac{\dot{Q}}{L} = \frac{2\pi k(T_i - T_e)}{\ln (r_e/r_i) + k/r_e h_e} \tag{3.18}$$

In actual cases, the prevailing conditions are far removed from that derived in Eq. (3.17). The effect of h_e is small in most cases so that an estimate of it proves to be adequate. Similarly, the effect of the metallic wall can, as a rule, be neglected.

Example 3.3 An oil duct of 100/108 mm in diameter is insulated with a layer of glass wool. The thickness of the layer is $s = 50$ mm. The thermal conductivity of the insulation encountered in practice (about 20% larger than its value obtained by measurement in the laboratory) is $k = 0.042$ W/m K. The temperature of the oil is $T_i = 50°$C and the temperature of the surroundings is $T_e = -15°$C. We assume $h_e = 30$ W/m² K (strong wind). Hence, Eq. (3.18) yields $\dot{Q}_L = 25.7$ W/m.

3.3.2 Temperature Drop in Pipes

Employing Eq. (3.18), we can set up a heat balance equation which governs the cooling of the fluid at an arbitrary location x, referring it to a length dx. Thus

$$-\dot{m}c_p \, dT_i = \dot{Q}_L \, dx = \frac{2\pi k(T_i - T_e) \, dx}{\ln (r_e/r_i) + k/r_e h_e} \tag{3.19}$$

where $\dot{m}$ denotes the mass-flow rate (SI unit: kg/s) and c_p is the specific heat at constant pressure of the fluid. Stipulating that the temperature at entry is T_1 at $x = 0$ and that the required temperature at exit $x = L$ is denoted by T_2, we integrate Eq. (3.19) and calculate that

$$\ln \frac{T_2 - T_e}{T_1 - T_e} = -\frac{L(\dot{Q}_L)_0}{\dot{m}c_p(T_1 - T_e)} \tag{3.20}$$

where $(\dot{Q}_L)_0$ denotes the value of $\dot{Q}_L$ at $x = 0$ and $T_1 = T_i$. If $T_1 - T_2 \ll T_1 - T_e$, we may linearize[†] Eq. (3.20) to obtain

$$T_1 - T_2 = \frac{L(\dot{Q}_L)_0}{\dot{m}c_p} \tag{3.21}$$

Example 3.4 We assume that the oil pipe discussed in Sec. 3.1.1 has a length of $L = 10$ km. If $\dot{m} = 6.4$ kg/s and $c_p = 2.1$ kJ/kg K and $\dot{Q}_L$ calculated earlier is identified with $(\dot{Q}_L)_0 = 25.7$ W/m at $x = 0$, we obtain

$$\ln \frac{T_2 - T_e}{T_1 - T_e} = -0.294$$

For $T_1 = 50°C$ and $T_e = -15°C$, we obtain $T_2 - T_e = 48.5$ K and $T_2 = 33.5°C \approx 34°C$, so that the loss of temperature is $T_1 - T_2 \approx 16$ K.

3.3.3 Cooling of Reservoirs

If the content of a reservoir has a uniform temperature T_i, and if it is permissible to neglect the thermal resistance of the container wall (subscript R) as well as the internal film coefficient, stipulating that

$$\frac{1}{h_i} + \frac{s_R}{k_R} \ll \frac{1}{h_e}$$

we can formulate the energy balance for a reservoir enclosing a mass m in a volume of area A as

$$-mc_p \, dT_i = h_e A(T_i - T_e) \, dt \tag{3.22}$$

In the case of lagged reservoirs, it is possible to replace h_e by K, where

$$\frac{1}{K} = \frac{1}{h_e} + \frac{s}{k}$$

Assuming $T_i = T_0$ for $t = 0$, we may integrate Eq. (3.22) as follows:

[†]$\ln(1-y) \approx -y$ where $y = (T_1 - T_2)/(T_1 - T_e)$.

$$\ln \frac{T_i - T_e}{T_0 - T_e} = -\frac{KAt}{mc_p} = -\frac{KAt}{V\rho c_p} = -\frac{t}{\tau} \qquad (3.23)$$

Here V is the volume of the container, ρ the density of its contents, and $\tau = mc_p/KA$ is the time constant of the reservoir. If $(T_0 - T_i) \ll (T_0 - T_e)$, we may linearize (3.23) to the form

$$T_0 - T_i = \frac{K(T_0 - T_e)At}{V\rho c_p} = (T_0 - T_e)\frac{t}{\tau} \qquad (3.24)$$

Example 3.5 We are asked to estimate the minimum diameter d that a spherical container must possess in order to reduce its initial temperature difference $T_0 - T_e$ by at most 20% in 6 months (storage of solar energy). With high-quality insulation we may assume $h_0 \gg k/s$ and hence $K \approx k/s$. For a sphere $A/V = 6/d$. Since $\ln 0.8 = -0.223$, and with $t = 6$ months $\approx 15.6 \times 10^6$ s, $s = 0.5$ m, $k = 0.03$ W/m K and $\rho c_p = 4180$ kJ/m^3 K (water), we get from Eq. (3.23) the result that

$$d \geqslant \frac{26.9kt}{s\rho c_p} \approx 6 \text{ m}$$

The preceding calculation convinces us that the problem of storing solar energy can be solved only on condition that very large reservoirs are employed. Placing the container underground would not change anything essential.

Example 3.6 A thermometer of temperature T_0 is immersed in a well-stirred bath of constant temperature $T_B \neq T_0$. According to Eq. (3.23), the thermometer lag is

$$T_B - T = (T_B - T_0) \exp\left(-\frac{t}{\tau}\right)$$

where $\tau = mc_p/KA$ is the time constant. German standards (PTB-Mitt., vol. 76, pp. 75-77, 1966) prescribe that $\tau \leqslant 2.6$ s for medical thermometers. If such a thermometer is immersed in a bath of 40°C from an initial state of 20°C, it will lag by 0.01 K after 20 s.

3.3.4 Thermometer in Heated Bath

We consider a temperature sensor whose indication is used to regulate the temperature T_B of a bath. The latter is heated according to the linear relation

$$T_B(t) = T_1 + bt \qquad (3.25)$$

We wish to calculate the error $T_B - T$, where T denotes the instantaneous temperature of the sensor. In the numerical evaluation we shall make use of $T_B(t)$ from Eq. (3.25) to replace T_e in Eq. (3.22). In the present problem, T corresponds to the previous T_i. This leads to the differential equation

$$\frac{\tau \, dT}{dt} = T_1 + bt - T \qquad (3.26)$$

where the time constant of the sensor is given by $\tau = mc_p/KA$. Assuming that the initial temperature of the sensor is T_0 $(T_0 \gtrless T_1)$, we arrive at the result that

$$T_B - T = (T_1 - T_0) \exp\left(-\frac{t}{\tau}\right) + b\tau \left[1 - \exp\left(-\frac{t}{\tau}\right)\right] \qquad (3.27)$$

For $b = 0$, with $T_1 = T_B$, we recover the conditions of Example 3.6 in Sec. 3.3.3. If $t \gg \tau$, we calculate that $T_B - T = b\tau$ is the residual indicator error. An example is illustrated with the aid of Fig. 3.4. Here $\tau = 5$ min, $b = 1$ K/min and the purposely chosen large temperature lag is $b\tau = 5$ K. It is seen that the initial large temperature differences $(T_0 > T_1)$ are made to decrease slowly with time and that all curves approach the common asymptote $T_B - b\tau$.

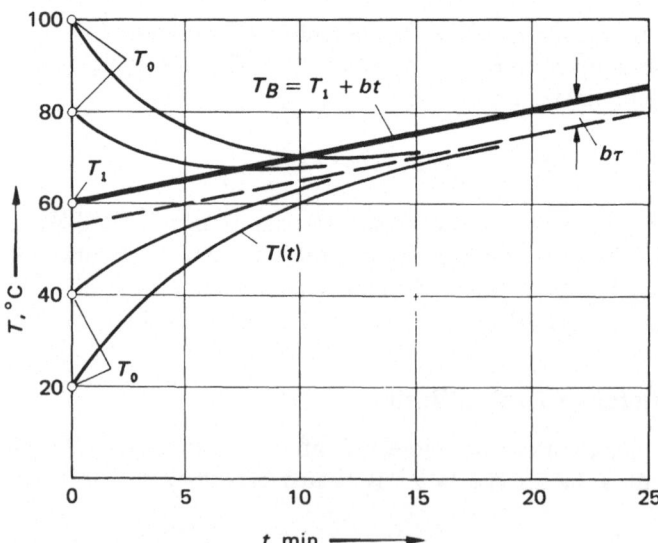

Figure 3.4 Temperature sensor in heated bath.

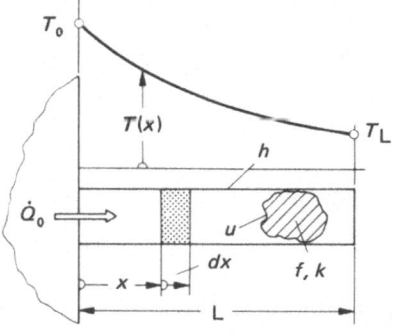

Figure 3.5 Illustrating the derivation of the temperature distribution along a rod or fin.

3.4 EXTENDED SURFACES

The heat transfer rate $\dot{Q}$ which crosses area A outward is given by Eq. (1.7) as

$$\dot{Q} = hA(T_w - T_\infty)$$

It is true that the film coefficient h can be increased, for example by increasing the flow velocity of the coolant, but this can be achieved only within narrow limits. In most applications the temperatures T_w and T_∞ are prescribed. Thus the only remaining possibility, if we wish to increase the heat transfer rate $\dot{Q}$, is to increase the surface A, for example by adding extensions in the form of ribs, fins, or pins, either of uniform or of tapered cross section. In this section we examine the question of whether and when the addition of such extended, or finned, surfaces serves our purpose and what is their effect on the heat transfer rate $\dot{Q}$.

We consider a rod or fin of length L of constant cross-sectional area f and circumference u over the length L, as shown in Fig. 3.5. The fin is connected to a wall at one end and transfers heat from it. The film coefficient at the circumference of the fin is h.

The temperature distribution $T(x)$ along the fin can be calculated by setting up a heat balance. This requires that the decrease in the heat transfer rate along the rod in the x-direction must be equal to the heat transfer rate dissipated through the circumference. Thus, for an element of length dx, we may write

$$\dot{Q}(x) - \dot{Q}(x + dx) = hT(x)u \, dx$$

when we put $T_\infty = 0$. Using the Taylor-series expansion

$$\dot{Q}(x + dx) = \dot{Q}(x) + \frac{d\dot{Q}}{dx} \, dx$$

and the Biot-Fourier law of conduction

$$Q = -kf \frac{dT}{dx}$$

we arrive at the following differential equation:

$$\frac{d^2 T}{dx^2} = \frac{hu}{kf} T = m^2 T \tag{3.28}$$

where the abbreviation

$$m = \left(\frac{hu}{kf}\right)^{1/2}$$

has been introduced (m has the dimension of reciprocal length). The general solution of Eq. (3.28) has the form

$$T = C_1 \exp(mx) + C_2 \exp(-mx) \tag{3.29}$$

whose constants of integration, C_1 and C_2, are to be determined to satisfy the boundary conditions. At $x = 0$ we must have $T = T_0$, where T_0 denotes the constant wall temperature. The quantity of heat which leaves the frontal area f at $x = L$ can be neglected for most purposes. This is due not only to the fact that f is small compared with uL, but also because the temperature distribution in that region has flattened considerably. Thus we put

$$\left(\frac{dT}{dx}\right)_L = 0$$

In view of this, we transform Eq. (3.29) to read

$$\frac{T}{T_0} = \frac{\exp[m(L-x)] + \exp[-m(L-x)]}{\exp(mL) + \exp(-mL)} = \frac{\cosh[mL(1-x/L)]}{\cosh mL} \tag{3.30}$$

The residual temperature excess at the end of the fin at $x = L$ is

$$T_L = \frac{T_0}{\cosh mL} \tag{3.31}$$

and the heat transfer rate at $x = 0$ becomes

$$\dot{Q}_0 = -kf\left(\frac{dT}{dx}\right)_0 = mkfT_0 \tanh mL \tag{3.32}$$

The temperature ratio T/T_0 is seen plotted against x/L in Fig. 3.6 with mL as a parameter. It is observed that the useful range is confined to the

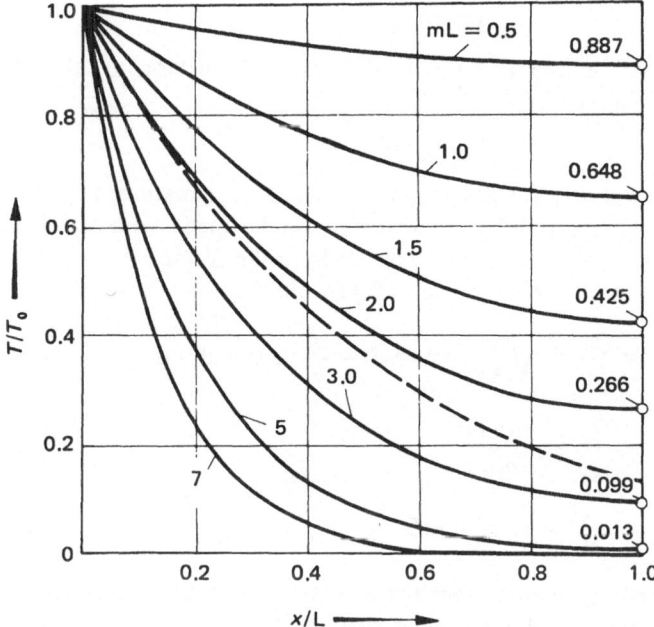

Figure 3.6 Temperature distribution T/T_0 along a fin in terms of the relative distance x/L, with the quantity mL as the parameter. The curves represent Eq. (3.30). The broken line represents Eq. (3.33) and corresponds to an infinite rod with $mL = 2$.

interval of $mL = 1$ to 1.5, roughly speaking. The fin with $mL = 0.5$ is too short because at its end the temperature excess equals about 89% of that at the root. The rod with $mL = 5$ or 7 is too long, because very little heat is dissipated over something like half of its length, owing to the considerable drop in excess temperature. In the case of an infinite fin with $T = 0$ at $x \to \infty$, Eq. (3.29) yields

$$\frac{T}{T_0} = \exp(-mx) = \exp\left(-mL \frac{x}{L}\right) \tag{3.33}$$

In this case, the heat transfer rate $\dot{Q}_0$ at $x = 0$ becomes

$$(\dot{Q}_0)_\infty = mkfT_0 \tag{3.34}$$

The function represented by Eq. (3.33) is shown as a broken line in Fig. 3.6, with $mL = 2$: it can be used as an approximation for fins of finite length which is valid for low values of x/L provided mL is not too small.

The heat transfer rate $\dot{Q}_0$ at $x = 0$ calculated from Eq. (3.32) can be compared with

$$\dot{Q}' = hfT_0$$

which would prevail in the absence of the extended surface, or, alternatively, with the quantity

$$\dot{Q}_k = hufT_0$$

which would exist if the fin developed a constant temperature, i.e., with that for $k \to \infty$. Alternatively, a comparison can be made with the heat transfer rate $(\dot{Q}_0)_\infty$ from Eq. (3.34). This leads to the formulation of three different figures of merit (or "effectiveness") for the fin:

$$\epsilon_w = \frac{\dot{Q}_0}{\dot{Q}'} = \sqrt{\frac{ku}{hf}} \tanh mL = \frac{k}{hL} mL \tanh mL \qquad (3.35)$$

$$\epsilon_k = \frac{\dot{Q}_0}{\dot{Q}_k} = \frac{\tanh mL}{mL} \qquad (3.36)$$

$$\epsilon_\infty = \frac{\dot{Q}_0}{(\dot{Q}_0)_\infty} = \tanh mL \qquad (3.37)$$

The figure of merit ϵ_w characterizes the increase (or, possibly, decrease) of the heat flow induced by the extended surface. The ratio ϵ_k characterizes the imperfection of the material (finite k), and ϵ_∞ describes the effect of

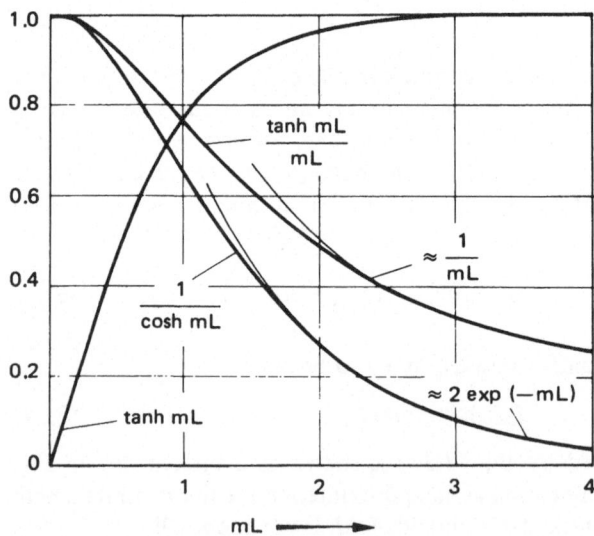

Figure 3.7 Hyperbolic functions which occur in the formulas for the calculation of the temperature distribution and the figures of merit in prismatic rods and fins.

the finite length. Several hyperbolic functions useful for the preceding calculations have been plotted in Fig. 3.7.

The proper design of an extended surface presents itself in the form of an optimization problem. It is, for example, possible to require that a given volume $V = fL$ of the material should be used optimally for the transfer of heat. In this case we substitute $f = V/L$, with the stipulation that $V = $ const, into Eq. (3.32) for the heat transfer rate $\dot{Q}_0$ which is to be rendered a maximum. It is further necessary to impose some condition on the relation between u and f, that is, in the last analysis on L. For this reason we investigate two variants which are important in practical applications.

Case a. If the shape of the cross-section f is to remain similar (as in the case of a circular cross section or a rectangular cross section of constant aspect ratio) in the course of the variation, we adopt the relation $u = C\sqrt{f}$. The explicit formula in the case of a circular cross section is here

$$C = 2\sqrt{\pi}$$

Its analog for a rectangular cross section is

$$C = \frac{2(n + 1)}{\sqrt{n}}$$

where $n = b/d$ is the ratio of the two sides. Applying the above supplementary condition, we can express $\dot{Q}_0$ from Eq. (3.32) as a unique function of the fin length L. This is

$$\dot{Q}_0 = \sqrt{hkC}\ V^{3/4}L^{-3/4}T_0 \tanh \left(\sqrt{\frac{hC}{kV^{1/2}}}\ L^{5/4} \right) \qquad (3.38)$$

Imposing the condition $d\dot{Q}/dL = 0$ for the existence of an extremum, we derive (after some calculation) the following transcendental determining equation:

$$\tanh mL = \frac{5mL}{3 \cosh^2 mL} \qquad (3.39)$$

Equation (3.39) is satisfied when

$$mL = \sqrt{\frac{hu}{kf}}\ L$$

assumes the numerical value 0.919. We omit the proof that the preceding extremum is indeed a maximum. In the case of this particular optimal design, the three figures of merit assume the numerical values: $\epsilon_w = 0.667 \; k/hL$, $\epsilon_k = 0.789$, $\epsilon_\infty = 0.725$. The temperature excess at the end of the fin ($x = L$) is $T_L = 0.688 \, T_0$.

Case b. Requirements of the process of manufacturing fins make it desirable to employ fins of rectangular cross section. In general, it is possible to assume that the width b of the fin is determined by the dimensions of the vessel that is to be cooled or heated. Thus, in addition to the length L it is necessary to include only the thickness d in the optimization calculation. We assume, further, that we must have $b/d \gg 1$ where $u \approx 2b$ and $f = bd$. Then Eq. (3.32) gives

$$\dot{Q}_0 = \sqrt{2b Vhk} \; T_0 \, L^{-1/2} \, \tanh \left(L^{3/2} \, \sqrt{\frac{2hb}{kV}} \right) \tag{3.40}$$

The imposition of the condition $d\dot{Q}_0 / dL = 0$ for the existence of an extremum leads to a transcendental equation which differs from that for Case a merely by a numerical constant, viz.,

$$\tanh mL = \frac{3mL}{\cosh^2 mL} \tag{3.41}$$

and which is satisfied by

$$mL = \sqrt{\frac{2h}{kd}} \, L = 1.419$$

The figures of merits are now: $\epsilon_w = 1.262 \; k/hL$, $\epsilon_k = 0.627$, $\epsilon_\infty = 0.889$, and the temperature excess at $x = L$ is $L = 0.457 \, T_0$.

In the special cases considered here, a measurement of the temperature excess at the end of the fin can give us an indication whether in a particular design optimum conditions have actually been attained.

Example 3.7 We consider the cylinder head of an aircraft engine which is provided with rectangular fins of optimum volume utilization. The fin width b, the film coefficient h, the temperature excess T_0, and the heat transfer rate $\dot{Q}_0$ to be dissipated are prescribed. The number n of fins and the metal (copper, aluminum, or steel) are to be determined so as to safeguard the smallest mass M. Assuming n fins, we compute, with the aid of Eq. (3.40), that

$$\dot{Q}_0 = n \, \sqrt{\frac{kV}{L}} \, T_0 \, \sqrt{2hb} \, \tanh 1.419$$

It follows from Eq. (3.41) that

Table 3.1 Comparison between different fin materials (at $T = 20°C$)

Metal	ρ, kg/m^3	k, W/m K	ρ/k, kg K/m^2 W
Cu	8300	372	22.3
Al	2300	142	16.2
Steel	7900	52	152

$$2h L^2 = (1.419)^2 \, kd$$

The mass of metal is

$$M = nV\rho = nbdL\rho$$

Eliminating V, d, and L, we obtain

$$\frac{\dot{Q}_0^3}{(0.889)^3 (T_0 \sqrt{2hb})^3} = \frac{kn^2 M}{\rho} \frac{\sqrt{2hb}}{1.419}$$

Finally,

$$M = \frac{1}{n^2}\left(\frac{\rho}{k}\right) \times 2.02 \frac{(\dot{Q}_0/T_0)^3}{(2hb)^2} = \frac{1}{n^2}\left(\frac{\rho}{k}\right) \times \text{const}$$

The preceding equation asserts that owing to the variability of $M \propto 1/n^2$ and $V \propto 1/n^3$, it is advisable to accommodate as many small fins as is technologically possible. In the selection of the material it is necessary to aim at achieving the smallest possible value of the ratio ρ/k. Table 3.1 shows that aluminum fins are preferable to those made of copper or steel.

FOUR

STEADY-STATE CONDUCTION IN THE PRESENCE OF HEAT SOURCES

In the examples analyzed so far, there existed heat sources and sinks in the system. They were indispensable for maintaining a flow of heat, but they were located outside the field under consideration. In what follows, we shall analyze cases in which heat sources are integral parts of the system. Examples include electric heaters, inductive or capacitance heating, chemical or nuclear reactions, the absorption of radiation, phase transformations, biological processes, and many others. In the steady-state case, and with constant values of k and $\dot{q}'$, Eq. (1.5) transforms into Poisson's equation

$$\nabla^2 T = \text{const} \tag{4.1}$$

For one-dimensional heat conduction through plates, cylinders, or spheres, we obtain

$$\frac{d^2 T}{dr^2} + \frac{n}{r} \frac{dT}{dr} + \frac{\dot{q}'}{k} = 0 \tag{4.2}$$

where r is the length coordinate measured from the center. The values of n for different configurations are listed in Table 4.1. Equation (4.2) can be solved with the aid of the substitution $dT/dr = u$, subject to the following boundary conditions: considerations of symmetry impose $dT/dr = 0$ in the

Table 4.1 Constants in the equations of one-dimensional conduction

	n	m	V/A
Plate	0	2	R
Cylinder	1	4	$\frac{1}{2}R$
Sphere	2	6	$\frac{1}{3}R$

center $(r = 0)$, whereas at the outer edge $r = R$ we must introduce the boundary condition of the third kind, namely,

$$-k\left(\frac{dT}{dr}\right)_R = h(T_w - T_\infty)$$

where T_∞ is the temperature of the fluid outside the thermal boundary layer. Thus, the solution to Eq. (4.2) assumes the form

$$T(r) - T_\infty = \frac{\dot{q}'R^2}{mk}\left(1 + \frac{2k}{hR} - \frac{r^2}{R^2}\right) \tag{4.3}$$

with $m = 2n + 2$ listed in Table 4.1.

According to Eq. (4.3), the temperature in the center, that is at $r = 0$, is given by

$$T_0 - T_\infty = \frac{\dot{q}'R^2}{mk}\left(1 + \frac{2k}{hR}\right) \tag{4.4}$$

The surface temperature T_w at $r = R$ is

$$T_w - T_\infty = \frac{2\dot{q}'R}{mh} \tag{4.5}$$

so that the difference $T_0 - T_w$ satisfies the equation

$$T_0 - T_w = \frac{\dot{q}'R^2}{mk} \tag{4.6}$$

In many cases, the internal source strength $\dot{q}'$ is not known, but the heat flux $\dot{q}_w$ at the wall of the region is prescribed. The energy balance dictates that we must have

$$\dot{q}'V = \dot{q}_w A \tag{4.7}$$

where V and A denote corresponding values of volume and surface. The ratios V/A are listed in Table 4.1. Employing these data, we are led to the unified equations for all three shapes, namely

$$T(r) - T_\infty = \frac{\dot{q}_w R}{2k} \left(1 + \frac{2k}{hR} - \frac{r^2}{R^2} \right) \tag{4.8}$$

$$T_0 - T_w = \frac{\dot{q}_w R}{2k} \tag{4.9}$$

$$T_w - T_\infty = \frac{\dot{q}_w}{h} \tag{4.10}$$

Example 4.1 In a nuclear reactor equipped with fuel elements in the form of rods of radius R and prescribed heat flux $\dot{q}_w$ at the outer surface, it is possible to calculate the excess temperature at the axis of the rod by the use of Eq. (4.9). Employing the numerical values of k indicated below, and for $\dot{q}_w = 100$ W/cm^2 with $R = 1$ cm, we calculate the following values of $T_0 - T_w$:

Fuel	k, W/m K	$T_0 - T_w$, K
Metallic uranium	29.7	168
Uranium carbide	36.8	136
Uranium dioxide	2.34	2140

The power output per unit length $\dot{q}_L = \dot{Q}/L$ (known as linear heat generation rate in nuclear engineering) can be calculated with the aid of Eq. (4.7) and Table 4.1, when we obtain

$$\dot{q}_L = \frac{\dot{Q}}{L} = \pi R^2 \dot{q}' = 2\pi \dot{q}_w R$$

We now eliminate $\dot{q}_w R$ from this equation with the aid of Eq. (4.9) and introduce the integral mean value k_m as well as the integral K from Sec. 5.2. This yields

$$\frac{\dot{q}_L}{4\pi} = k_m(T_0 - T_w) = \int_{T_w}^{T_0} k(T)\, dT = K_{wo} \tag{4.11}$$

This relation allows us to calculate the integral thermal conductivity K from a measurement of the linear heat generation rate $\dot{q}_L$ alone.

In order to lay down the two temperature limits in the integral, we must know the two temperatures T_0 and T_w. At a minimum, we must be able to estimate them. In the previous example, we have $\dot{q}_L = 628$ W/cm, and the power density referred to the volume of the pure fuel element, i.e., not counting additional structural parts, is $\dot{q}' = 200$ W/cm^3. The integral thermal conductivity is $K = 5000$ W/m.

FIVE

STEADY-STATE CONDUCTION IN SEVERAL DIMENSIONS

Steady-state conduction in several dimensions and in the absence of internal heat sources is governed by the Laplace equation

$$\nabla^2 T = 0 \tag{5.1}$$

which, written in rectangular coordinates, assumes the form

$$\frac{\partial^2 T}{\partial x^2} + \frac{\partial^2 T}{\partial y^2} + \frac{\partial^2 T}{\partial z^2} = 0 \tag{5.2}$$

The solutions of Eqs. (5.1) and (5.2) form systems of isothermal surfaces which it is possible to supplement with systems of orthogonal adiabatic surfaces. In most cases, at the boundaries of the region under consideration, we are given the temperatures (boundary conditions of the first kind). As a rule we wish to calculate the heat flux that passes between two or more isothermal surfaces. These may or may not be closed. If they are not closed, the remaining boundaries of the system are adiabatic. One of the isothermal boundaries can also occur at infinity. In two-dimensional problems, the boundaries consist of lines instead of surfaces.

5.1 SHAPE FACTOR AND SHAPE RESISTANCE

The heat transfer rate $\dot{Q}$ from isothermal surface A_1 to isothermal surface A_2 at temperatures T_1 and T_2, respectively, can be obtained by the integration of local temperature gradients $\partial T/\partial n$ over the surfaces as follows:

$$\dot{Q} = -k \iint_{A_1} \left(\frac{\partial T}{\partial n}\right)_1 dA_1 = k \iint_{A_2} \left(\frac{\partial T}{\partial n}\right)_2 dA_2 \qquad (5.3)$$

where $\partial/\partial n$ denotes differentiation in the normal direction (counted positive when pointing inward). The heat transfer rate can also be written as

$$\dot{Q} = kS(T_1 - T_2) \qquad (5.4)$$

which introduces the shape factor S. Equations (5.3) and (5.4) provide for S the defining relations

$$S = \frac{\iint_{A_1} (\partial T/\partial n)_1 \, dA_1}{T_2 - T_1} = -\frac{\iint_{A_2} (dT/dn)_2 \, dA_2}{T_2 - T_1} \qquad (5.5)$$

The shape factor constitutes a geometrical quantity and its dimension is that of length; it is independent of the thermal conductivity k as well as of the temperature difference $T_2 - T_1$.

For bodies accommodating two-dimensional temperature distributions whose length at right angles to the plane of the temperature distribution is L, it is convenient to introduce the specific shape factor $S_L = S/L$; the latter is dimensionless.

Example 5.1 The shape factors S and S_L of an insulating shell, made in the form of a hollow cylinder whose radii are r_e and r_i, Eq. (3.8), are given by the expressions

$$S = \frac{2\pi L}{\ln(r_e/r_i)} \qquad S_L = \frac{2\pi}{\ln(r_e/r_i)}$$

The shape factors must be added together in cases when heat flows are arranged in parallel. Thus, for an insulator made in the form of half of a hollow cylinder, we may write $S_L = \pi/\ln(r_e/r_i)$.

When the heat flows are arranged in series, it is necessary to add the reciprocals of the shape factors; that is, form resistances $R = 1/S$ or $R_L = 1/S_L$ must be added together. If we were to add a second insulating layer whose radii are r_b and r_e $(r_b > r_e)$ to the shell discussed in the preceding example, we would get

$$\frac{1}{S_L} = R_L = \frac{\ln(r_e/r_i) + \ln(r_b/r_e)}{2\pi} = \frac{\ln(r_b/r_i)}{2\pi}$$

on the stipulation that the joint along the common cylindrical surface offers no additional resistance to heat flow.

The concept of a shape factor can be usefully applied to other physical processes which are governed by the Laplace equation. The charge Z on an electric condenser is

$$Z = \epsilon S(V_1 - V_2) \tag{5.6}$$

where ϵ is the dielectric constant (SI unit: As/Vm) and $V_1 - V_2$ denotes the potential difference between the condenser plates.

The steady-state electric current i in a homogeneous space is

$$i = \sigma S(V_1 - V_2) \tag{5.7}$$

the symbol σ denoting the electric conductance (SI unit: A/Vm). In classes of systems which preserve geometric similarity, all shape factors S in the above three cases acquire the same numerical values. The preceding property forms the basis of the operation of the electrolytic tank discussed in Sec. 5.7.1.

5.2 KIRCHHOFF'S TRANSFORMATION

The equations of the preceding sections have been derived for homogeneous and isotropic bodies whose thermal conductivity does not depend on temperature. It is possible to extend the use of all results pertaining to steady-state conduction to include temperature-dependent thermal conductivity $k(T)$ by the use of a transformation first introduced into science by Kirchhoff [5.1]. We start with Eq. (1.3) retaining a heat source in it, that is, with

$$\frac{\partial}{\partial x}\left(k\,\frac{\partial T}{\partial x}\right) + \frac{\partial}{\partial y}\left(k\,\frac{\partial T}{\partial y}\right) + \frac{\partial}{\partial z}\left(k\,\frac{\partial T}{\partial z}\right) + \dot{q}' = 0 \tag{5.8}$$

Kirchhoff introduced the new temperature Θ through the definition

$$k_m\,d\Theta = k\,dT \tag{5.9}$$

in which $k(T)$ is the temperature-dependent thermal conductivity and k_m denotes a constant mean value. Introducing Eq. (5.9) into Eq. (5.8), we get

$$\frac{\partial^2 \Theta}{\partial x^2} + \frac{\partial^2 \Theta}{\partial y^2} + \frac{\partial^2 \Theta}{\partial z^2} + \frac{\dot{q}'}{k_m} = 0 \tag{5.10}$$

This transformed equation shows how all results obtained for $k = \text{const}$ can be extended to the case $k = k(T)$. It is usually most convenient so to choose the new temperature Θ as to make it equal to the highest and lowest natural temperatures of the problem, respectively, i.e., to stipulate $\Theta_2 = T_2$ and $\Theta_1 = T_1$. Integrating Eq. (5.9)–paying attention to distinguish between the cases when $T_2 > T_1$ and when $T_1 > T_2$–we obtain

$$k_m(\Theta_2 - \Theta_1) = k_m(T_2 - T_1) = \int_{T_1}^{T_2} k(T)\, dt \qquad \text{for } T_2 > T_1$$

and $\quad k_m(\Theta_1 - \Theta_2) = k_m(T_1 - T_2) = \displaystyle\int_{T_2}^{T_1} k(T)\, dt \qquad \text{for } T_1 > T_2$

Hence the mean value is defined by

$$k_m = \frac{1}{T_2 - T_1} \int_{T_1}^{T_2} k(T)\, dt \qquad \text{or} \qquad k_m = \frac{1}{T_1 - T_2} \int_{T_2}^{T_1} k(T)\, dT$$

$$(5.11)$$

If the function $k(T)$ is a linear one, the value of k_m corresponds to the mean temperature $\frac{1}{2}(T_1 + T_2)$. This approximation suffices for many practical applications. The temperatures at the boundaries are not known beforehand when boundary conditions of the third kind apply. In such cases an estimate is made first and improved by iteration.

In the case of systems which sustain very large temperature differences (such as nuclear fuel rods), it is useful to tabulate not only the values of the function $k(T)$, but also of the integral thermal conductivities

$$K_{12} = \int_{T_1}^{T_2} k(T)\, dT \qquad \text{or} \qquad K_{21} = \int_{T_2}^{T_1} k(T)\, dt$$

with respect to an arbitrarily chosen initial temperature, for example, $0°C$. The required value K_{12} can be obtained from the tabulated set $K_{0T}(T)$ with the aid of the obvious relation

$$K_{12} = K_{02} - K_{01}$$

Table 5.1 lists by way of example the values of $k(T)$ and K for uranium dioxide from $0°$ to $2800°C$ (melting point) reproduced from [5.2]. These results were made use of earlier in Eq. (4.11). We now write the equation for the heat transfer rate $\dot{Q}$, introducing the shape factor S according to Eq. (5.4) and omitting the heat sources. With k_m from Eq. (5.11) we obtain

$$\dot{Q} = -k_m S(T_2 - T_1) = -S \int_{T_1}^{T_2} k(T)\, dT = -SK_{12} \qquad \text{for } T_2 > T_1$$

$$(5.12a)$$

Table 5.1 Thermal conductivity k and integral thermal conductivity K for uranium dioxide in terms of the Celsius temperature [5.2]

T, °C	k, W/m K	K_{0T}, W/m
0	10.73	0
100	8.24	937
500	4.30	3270
1000	2.81	4980
1500	2.35	6250
2000	2.39	7420
2500	2.84	8710
2800	3.29	9630

and $\dot{Q} = k_m S(T_1 - T_2) = S \displaystyle\int_{T_2}^{T_1} k(T)\, dT = SK_{21}$ for $T_1 > T_2$

$$(5.12b)$$

The heat transfer rate $\dot{Q}$, between the isothermal surfaces maintained at temperatures T_1 and T_2, turns out simply to be equal to the product of shape factor and integral thermal conductivity, K_{12} or K_{21}, as the case may be.

5.3 CONFORMAL MAPPING

In what follows we shall analyze two-dimensional stationary temperature fields without heat sources and with constant thermal conductivity. We begin by applying the condition of continuity of heat flux, div $\dot{q} = 0$, discussed in Chapter 1, and write

$$\frac{\partial \dot{q}_x}{\partial x} + \frac{\partial \dot{q}_y}{\partial y} = 0 \tag{5.13}$$

Introducing, further, the Biot-Fourier law

$$\dot{q}_x = -k\frac{\partial T}{\partial x} \quad \text{and} \quad \dot{q}_y = -k\frac{\partial T}{\partial y} \tag{5.14}$$

we derive Laplace's equation

$$\frac{\partial^2 T}{\partial x^2} + \frac{\partial^2 T}{\partial y^2} = 0 \tag{5.15}$$

Differentiating the auxiliary equation (5.14) with respect to y and x, in that order, we establish the condition

$$\frac{\partial \dot{q}_x}{\partial y} - \frac{\partial \dot{q}_y}{\partial x} = 0 \qquad (5.16)$$

or, in a more general form

$$\text{curl } \dot{q} = 0$$

which proves that the vector field of $\dot{q}$ is irrotational.

Laplace's equation (5.15) is satisfied not only by the scalar function $T(x, y)$, which we can assume to have been defined by Eq. (5.14), but also by another scalar function, $\psi(x, y)$, the so-called stream function, which is defined by the relations

$$\dot{q}_x = -k\frac{\partial \psi}{\partial y} \qquad \text{and} \qquad \dot{q}_y = k\frac{\partial \psi}{\partial x} \qquad (5.17)$$

Insertion of these expressions into Eq. (5.16) yields

$$\frac{\partial^2 \psi}{\partial x^2} + \frac{\partial^2 \psi}{\partial y^2} = 0 \qquad (5.18)$$

Comparing Eqs. (5.14) and (5.17), we derive that

$$\frac{\partial T}{\partial x} = \frac{\partial \psi}{\partial y} \qquad \text{and} \qquad \frac{\partial T}{\partial y} = -\frac{\partial \psi}{\partial x} \qquad (5.19)$$

These relations are known as the Cauchy-Riemann conditions; they follow directly from the definitions of T and ψ and are equivalent to the statement that both satisfy the Laplace equation.

The Cauchy-Riemann conditions provide the proof that the two families of curves, namely $T = \text{const}$ and $\psi = \text{const}$, form an orthogonal network. This can be seen by considering the two directional tangents

$$\left(\frac{dy}{dx}\right)_{T=\text{const}} = \frac{\partial T/\partial x}{\partial T/\partial y} = \frac{\dot{q}_x}{\dot{q}_y} \qquad (5.20)$$

$$\left(\frac{dy}{dx}\right)_{\psi=\text{const}} = \frac{\partial \psi/\partial x}{\partial \psi/\partial y} = -\frac{\dot{q}_y}{\dot{q}_x} \qquad (5.21)$$

Since both functions $T(x, y)$ and $\psi(x, y)$ satisfy the Laplace equation, it is possible to interchange their physical interpretation; this property of interchangeability extends also to the boundary conditions. We introduce the operator $\partial/\partial n$ to describe differentiation in a direction normal to an isotherm, that is in the direction of a thermal stream-line, and the operator $\partial/\partial s$ to denote differentiation along an isotherm. The Cauchy-Riemann conditions then lead to the conclusion that

$$\frac{\partial T}{\partial n} = \frac{\partial \psi}{\partial s} \qquad (5.22)$$

because

$$\frac{\partial T}{\partial n} = \frac{\partial T}{\partial x} \cos \phi + \frac{\partial T}{\partial y} \sin \phi = \frac{\partial \psi}{\partial y} \cos \phi - \frac{\partial \psi}{\partial x} \sin \phi = \frac{\partial \psi}{\partial s}$$

where ϕ is the angle included between the x and n directions. The heat transfer rate $\dot{Q}$, which crosses an isotherm between points s_1 and s_2 is

$$\dot{Q} = -kL \int_{s_1}^{s_2} \frac{\partial T}{\partial n} \, ds = -kL \int_{s_1}^{s_2} \frac{\partial \psi}{\partial s} \, ds = kL(\psi_1 - \psi_2) \quad (5.23)$$

The height of the body under consideration has here been denoted by L, and ψ_1, ψ_2 denote the values of the stream functions at locations s_1 and s_2 on the isotherm. It is convenient to rewrite these equations by introducing into them the reduced shape factor $S_L = S/L$ from Sec. 5.1 and in analogy with Eq. (5.4). Thus

$$\dot{Q} = kL\, S_L(T_1 - T_2) \qquad (5.24)$$

which Eq. (5.23) shows to be equivalent to

$$S_L = \frac{\psi_1 - \psi_2}{T_1 - T_2} \qquad (5.25)$$

The reduced shape factor S_L is, indeed, dimensionless, because according to our definition in Eq. (5.17), the stream function ψ has the dimension of temperature. If stream lines and isotherms are interchanged in their interpretation, the corresponding value of the reduced shape factor becomes the reciprocal of that in Eq. (5.25). This property is valid only for two-dimensional temperature fields.

The method of conformal mapping, [5.3], exploits the fact that certain functions[†] of a complex variable $z = x + iy$ satisfy the Cauchy-Riemann conditions. We assume that $w = f(z)$ is such a function and represent it as

$$w = f(z) = f(x + iy) = u(x, y) + iv(x, y)$$

where u and v are two real functions of the real variables x and y. The function $w = f(x)$ must be differentiable at point z which is equivalent to asserting that the limit

---

[†]So-called analytic functions.

$$\lim_{\epsilon \to 0} \frac{f(z + \epsilon) - f(z)}{\epsilon} = f'(z)$$

exists. Here $\epsilon = h + ik$ is a complex number which vanishes along an arbitrary path, such as along $x = $ const or $y = $ const. This means that the partial derivatives with respect to x and iy of

$$w(x + iy) = u(x, y) + iv(x, y)$$

must be equal, because we must have

$$\frac{\partial w}{\partial x} = \frac{\partial u}{\partial x} + i \frac{\partial v}{\partial x} = \frac{1}{i} \frac{\partial w}{\partial y} = \frac{1}{i} \frac{\partial u}{\partial y} + \frac{\partial v}{\partial y}$$

Comparing the real and imaginary parts, we derive the Cauchy-Riemann conditions

$$\frac{\partial u}{\partial x} = \frac{\partial v}{\partial y} \qquad \text{and} \qquad \frac{\partial u}{\partial y} = - \frac{\partial v}{\partial x}$$

Functions which satisfy these conditions are known as analytic functions. We have shown earlier that such functions satisfy the Laplace equation. This fact leads to the formulation of the theorem which is fundamental to the method of conformal mapping.

Theorem Every analytic function $w(z)$ of a complex variable $z = x + iy$ is a solution of the Laplace equation. Such a function defines two mutually orthogonal families of curves, $u(x, y)$ and $v(x, y)$, which can interchangeably be interpreted as isotherms or thermal stream lines.

The function $z = x + iy$ is also a solution of the Laplace equation. The lines $x = $ const and $y = $ const can be interpreted as isotherms and thermal stream lines. In this case they represent the one-dimensional conduction of heat across a flat plate. Similarly, the standard form $z = r \exp (i\phi)$ in cylindrical polar coordinates (r, radius vector; ϕ, polar angle) is also a solution of the Laplace equation. The circles $r = $ const are isotherms, and the radii $\phi = $ const are thermal stream lines. The function represents one-dimensional flow through a cylindrical shell.

The central problem of conformal mapping consists in a search for suitable analytic functions $w(z)$ with the aid of which we can reduce cases characterized by complicated boundaries to one or the other of the two one-dimensional patterns just mentioned. Frequently, success does not come until several successive transformations have been identified and performed. We now illustrate this method with a simple example.

We are given the function $w = z^2$ which leads to

$$w = u + iv = z^2 = (x + iy)^2 = x^2 + i2xy - y^2$$

Comparing the real and imaginary parts, we perform the identifications

$$u = x^2 - y^2 \qquad \text{and} \qquad v = 2xy$$

The straight lines $u = $ const and $v = $ const of the w-plane transform into hyperbolas in the z plane. In order to verify that z^2 is an analytic function, we calculate

$$\frac{\partial u}{\partial x} = 2x \qquad \frac{\partial v}{\partial y} = 2x \qquad \frac{\partial u}{\partial y} = -2y \qquad \frac{\partial v}{\partial x} = 2y$$

Thus we verify that the Cauchy-Riemann conditions are satisfied.[†] The distortion introduced by the transformation is recognized most easily by the juxtaposition of the two standard forms: $z = r \exp (i\phi)$ and $w = \rho \exp (i\omega)$, where ρ, ω denote, respectively, the radius vector and the polar angle in the w plane. Since $z^2 = r^2 \exp (i2\phi)$, we see that

$$\rho = r^2 \qquad \text{and} \qquad \omega = 2\phi$$

The radii of the w plane are squares of those of the z plane, and the polar angles in the w plane are doubles of those in the z plane. The preceding mapping is shown in Fig. 5.1a and b which represent the two planes. The two segments of the u axis (along $v = 0$) in the w plane located, respectively, between 0 and u_1 or $-u_1$ are turned by an angle $\frac{1}{2}\pi$ and place themselves along the x and y axes of plane z. The straight line $v = v_1$ has transformed into an equilateral hyperbola and the positive v axis (along $u = 0$) becomes the bisector in the first quadrant of the z plane. The complete figure is

[†]This would be evident for a student well-versed in the theory of analytic functions.

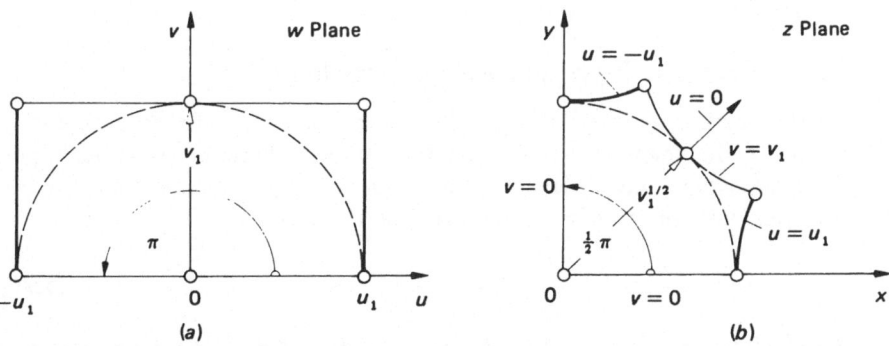

Figure 5.1a and b Conformal mapping of the function $w = z^2$.

closed by two hyperbolas that correspond to the straight lines $u = u_1$ and $u = -u_1$ of the w plane.

We can interpret the rectangle of the w plane as a section of a flat plate which sustains a one-dimensional flow of heat in the v direction. If this is to be so, then the boundaries $v = 0$ and $v = v_1$ must correspond to the isotherms T_1 and T_2. The heat transfer rate for a thickness L of plate measured at right angles to the plane of the drawing is, thus,

$$\dot{Q}_L = \frac{\dot{Q}}{L} = k \frac{2u_1}{v_1} (T_1 - T_2)$$

The reduced shape factor of this arrangement turns out to be equal to

$$S_L = \frac{S}{L} = \frac{2u_1}{v_1}$$

We obtain the same heat transfer rate, and the same reduced shape factor in the z plane, except that it now applies to a body confined within a different physical boundary. The two segments $v = 0$ on the x and y axis and the single equilateral hyperbola $v = v_1$ now play the role of the two isotherms.

Analogous statements apply to the heat transfer rate in the w plane in the u direction. The reduced heat transfer rate is

$$\dot{Q}'_L = k \frac{v_1}{2u_1} (T_1 - T_2)$$

because the straight lines $u = -u_1$ and $u = u_1$ have become the isotherms T_1 and T_2. The reduced shape factor is now

$$S'_L = \frac{v_1}{2u_1}$$

which retains its value upon transfer to the z plane. Comparison shows that $S_L = 1/S'_L$.

5.3.1 Eccentric Pipes and Related Problems

In this section we propose to employ the method of conformal mapping to calculate the shape factor for the flow of heat between eccentric pipes, between a pipe and a plane, and between two pipes in an extended space. For this purpose, we employ the mapping function

$$w = u + iv = \frac{z + if}{z - if} \tag{5.26}$$

in which $z = x + iy$ and f is a real number. Equation (5.26) is a special case of the general linear transformation

$$w = \frac{az + ib}{cz + id}$$

in which a, b, c, and d are complex numbers. This linear mapping function transforms circles into circles in which straight lines, that is circles of infinitely large radius, are also included. In order to separate the real and imaginary parts in Eq. (5.26), it is necessary to multiply the numerator and denominator by the complex quantity $x - iy + if$ which is conjugate to the denominator. After some calculation, we arrive at the result that

$$u = \frac{x^2 + y^2 - f^2}{x^2 + (y - f)^2} \quad \text{and} \quad v = \frac{2fx}{x^2 + (y - f)^2} \quad (5.27)$$

A circle of radius ρ centered at the origin in the w plane is represented by the equation

$$u^2 + v^2 = \rho^2 \quad (5.28)$$

We now substitute in Eq. (5.28) u and v from Eq. (5.27) and obtain in the z plane the geometrical image which corresponds to the circle in the w plane according to the mapping function (5.26). Introducing the reduced coordinates x/f and y/f in the z plane, we derive the relation

$$\left(\frac{x}{f}\right)^2 + \left(\frac{y}{f} - \frac{\rho^2 + 1}{\rho^2 - 1}\right)^2 = \frac{4\rho^2}{(\rho^2 - 1)^2} \quad (5.29)$$

This is the equation of a circle in the z plane whose radius is r/f and whose center has been displaced by a distance b/f in the direction of the positive y/f axis. From Eq. (5.29) we can deduce the relations

$$\frac{b}{f} = \frac{\rho^2 + 1}{\rho^2 - 1} \qquad \frac{r}{f} = \frac{2\rho}{|\rho^2 - 1|} \qquad \left|\frac{b}{r}\right| = \frac{\rho^2 + 1}{\rho^2} \quad (5.30)$$

The radius ρ of the w plane is thus

$$\rho = \left|\frac{b}{r}\right| \pm \sqrt{\frac{b^2}{r^2} - 1} \quad (5.31)$$

Since Eq. (5.30) shows also that

$$f = (b^2 - r^2)^{1/2}$$

we conclude that $b = f$ at $r = 0$.

The diagram in Fig. 5.2 depicts the transformation. The concentric circles of radius ρ of the w plane have been transformed into two sets of circles, each with a common source point, in the z plane. The source centers of these two sets of circles, as distinct from their geometric centers, are placed at $y/f = \pm 1$. The circle of radius $\rho = 1$ corresponds to the x axis

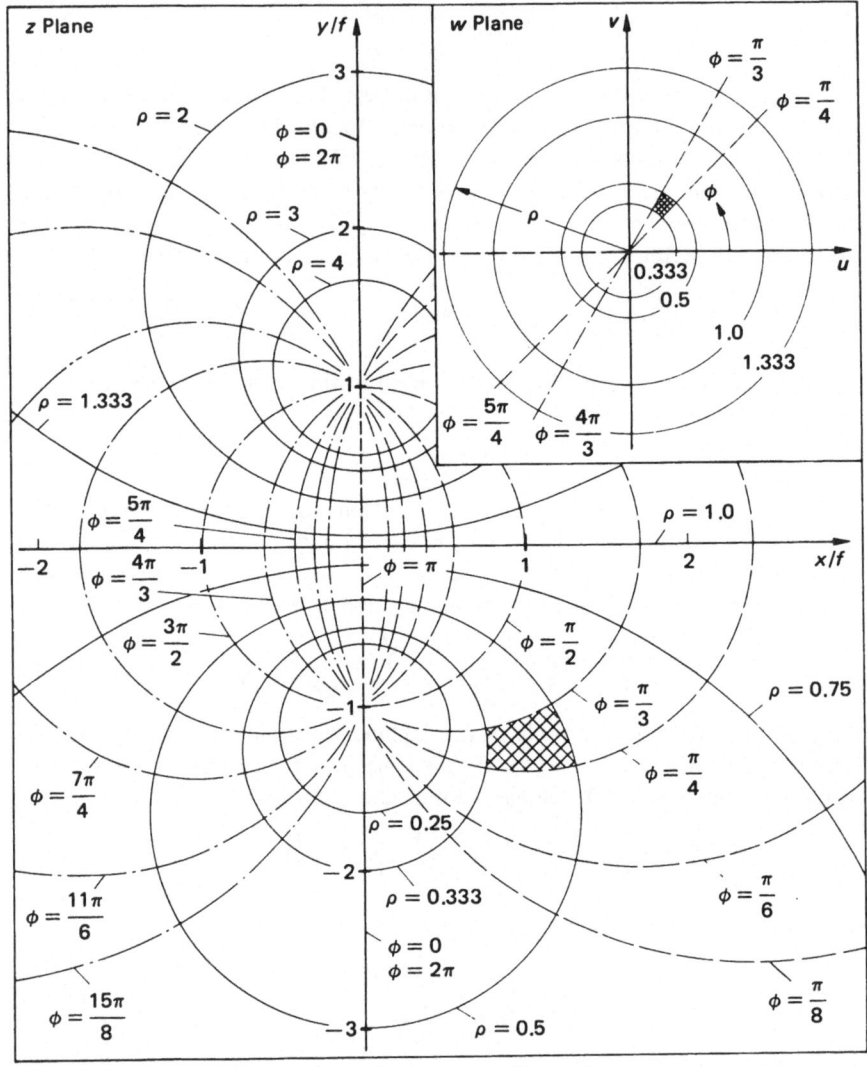

Figure 5.2 Linear mapping between the planes w and z.

of the z plane. All circles with $\rho > 1$ are placed in the upper half-plane and those for $\rho < 1$ are placed in the lower half-plane. The reduced heat transfer rate between two circles ($\rho_2 > \rho_1$) in the w plane is

$$\dot{Q}_L = \frac{2\pi k}{\ln(\rho_2/\rho_1)}(T_1 - T_2) \tag{5.32}$$

its direction is from ρ_1 to ρ_2. The reduced shape factor is now

$$S_L = \frac{2\pi}{\ln\,(\rho_2/\rho_1)} \qquad (5.33)$$

The expressions for the reduced heat transfer rate $\dot{Q}_L$, and the reduced shape factor S_L retain their validity for the associated pattern in the z plane if ρ from Eq. (5.31) is substituted. For ease of calculation, we recall that

$$\ln\,(x + \sqrt{x^2 - 1}\,) = \cosh^{-1} x$$

and $\qquad \cosh^{-1} x \pm \cosh^{-1} y = \cosh^{-1}\,(xy \pm \sqrt{(x^2 - 1)(y^2 - 1)}\,)$

Depending on the choice of ρ_1 and ρ_2, we derive the following shape factors.

Case 1. Two eccentric pipes, $\rho_2 < 1$ and $\rho_1 < \rho_2$, as in Fig. 5.3. Alternatively, $\rho_1 > 1$ and $\rho_2 > \rho_1$. In this case

$$S_L = \frac{2\pi}{\ln\,(\rho_2/\rho_1)} = \frac{2\pi}{\cosh^{-1}\,[(r_1^2 + r_2^2 - e^2)/2r_1 r_2]} \qquad (5.34)$$

where r_1, r_2 denote the pipe radii, and $e = |b_2 - b_1|$ is the eccentricity (distance between pipe center lines).

Case 2. Two pipes in extended space, $\rho_2 > 1$ and $\rho_1 < 1$, as in Fig. 5.4. Here

$$S_L = \frac{2\pi}{\ln\,(\rho_2/\rho_1)} = \frac{2\pi}{\cosh^{-1}\,[(d^2 - r_1^2 - r_2^2)/2r_1 r_2]} \qquad (5.35)$$

with r_1, r_2 denoting the radii, and $d = |b_1| + b_2$ representing the distance between the pipe center lines.

Case 3. Two pipes of equal diameters in extended space, $\rho_2 = 1/\rho_1 = \rho > 1$ with $b = |b_1| = b_2$. In this case

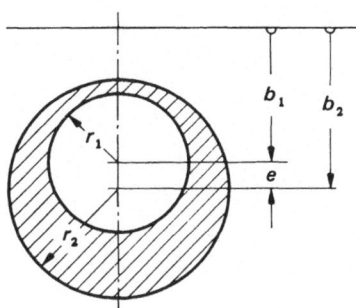

Figure 5.3 Eccentric pipes.

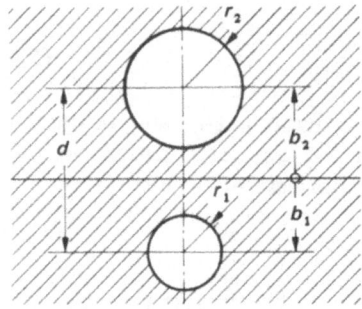

Figure 5.4 Two pipes in extended space.

$$S_L = \frac{2\pi}{2 \ln \rho} = \frac{\pi}{\cosh^{-1} (b/r)} = \frac{\pi}{\cosh^{-1} (d/2r)} \qquad (5.36)$$

This result follows from that in Case 2 with the substitution $d = 2b$ and $r = r_1 = r_2$.

Case 4. Pipe of radius r buried in ground at a depth $t = |b_1|$, $\rho_2 = 1$, $\rho = 1/\rho_1 > 1$, as in Fig. 5.5. Now

$$S_L = \frac{2\pi}{\ln \rho} = \frac{2\pi}{\cosh^{-1} (t/r)} \qquad (5.37)$$

The reduced shape factor S_L given in Eq. (5.37) is twice S_L from Eq. (5.36), because Case 3 pertains to two equal thermal resistances connected in series, as compared with the single resistance in Case 4.

In order further to facilitate calculations with Eq. (5.34)–(5.37), it is possible to employ the approximations

$$\cosh^{-1} x \approx \ln (2x) \qquad \text{for } x > 2.6 \qquad (5.38)$$

$$\cosh^{-1} x \approx [2(x - 1)]^{1/2} \qquad \text{for } x < 1.3 \qquad (5.39)$$

with an error $<2.5\%$

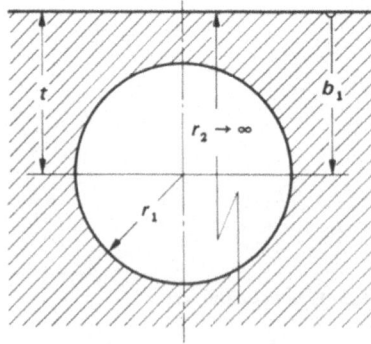

Figure 5.5 Buried pipe.

Example 5.2 A pipe of radius $r = 250$ mm is buried in sand ($k = 1.5$ W/m K), its center line being placed at a depth $t = 750$ mm below the surface. The reduced shape factor is

$$S_L = \frac{2\pi}{\cosh^{-1}(t/r)} = \frac{2\pi}{\cosh^{-1} 3} = 3.56$$

and the heat loss from a pipe of temperature $T_1 = 100°C$ in the presence of a temperature $T_2 = 0°C$ at the surface is

$$\dot{Q}_L = kS_L(T_1 - T_2) = 1.5 \text{ W/m K} \times 3.56 \times 100 \text{ K} = 534 \text{ W/m}$$

Conformal mapping furnishes exact values for the heat transfer rate only between two isotherms, that is for problems with boundary conditions of the first kind. Problems involving finite heat transfer rates (boundary conditions of the third kind) can be solved approximately with the aid of the method of auxiliary walls.

Example 5.3 We extend the conditions of Example 5.2 to stipulate that the air above the ground has a temperature of $0°C$ with a film coefficient $h = 10$ W/m² K. These data allow us to calculate the thickness of a fictitious depth t^*, given as

$$t^* = t + \frac{k}{h} = \left(0.75 + \frac{1.5}{10}\right) \text{m} = 0.9 \text{ m}$$

The reduced shape factor is now $S_L = 3.21$, and the heat loss becomes equal to $\dot{Q}_L = 428$ W/m. In this case the earth's surface ceases to be an isotherm.

5.4 FICTITIOUS HEAT SOURCES AND SINKS

Since the Laplace equation is linear, it follows that sums of solutions are also solutions.[†] Thus it is possible to simulate a given field of isotherms by a superposition of real or fictitious heat sources or sinks. As an example, we consider the case of a buried pipe or cable already familiar to us from Sec. 5.3.1. In accordance with Fig. 5.6, we introduce a cylindrical heat source of radius r_0 and temperature T_0 with its center at a distance f from the surface ($r_0 \ll f$). In order to generate the isotherms in the ground, whose surface temperature is assumed to be $T = 0$, we imagine a cylindrical sink of radius r_0 and temperature $-T_0$ placed at the mirror image of the

---

[†]More generally, linear combinations of solutions are also solutions.

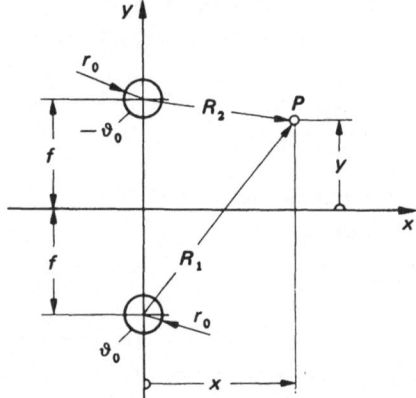

Figure 5.6 Pipe in ground with fictitious heat sink.

source. The sink absorbs exactly the heat transferred by the source. These heat transfer rates can be written as

$$\dot{Q}_{L1} = \frac{2\pi k(T_0 - T_a)}{\ln (R_1/r_0)}$$

and

$$\dot{Q}_{L2} = \frac{2\pi k[T_b - (-T_0)]}{\ln (R_2/r_0)}$$

Each of these generates at point P the temperatures T_a and T_b and their joint effect is to induce there the temperature

$$T = T_a + T_b = \frac{\dot{Q}_L \ln (R_2/R_1)}{2\pi k} = \frac{\dot{Q}_L \ln (R_2^2/R_1^2)}{4\pi k} \qquad (5.40)$$

because we postulate that $\dot{Q}_{L1} = \dot{Q}_{L2} = \dot{Q}_L$.

We now substitute Eq. (5.40) into the two relations, which can be easily deduced from Fig. 5.6. These are

$$R_1^2 = x^2 + (f + y)^2 \qquad \text{and} \qquad R_2^2 = x^2 + (f - y)^2$$

and they lead to

$$T = \frac{\dot{Q}_L}{4\pi k} \ln \frac{x^2 + (f - y)^2}{x^2 + (f + y)^2}$$

or

$$\exp \frac{4\pi k T}{\dot{Q}_L} = \frac{x^2 + (f - y)^2}{x^2 + (f + y)^2} = \kappa^2 = \frac{R_2^2}{R_1^2} \qquad (5.41)$$

It follows that the isotherms $T = \text{const}$, and hence the lines $\kappa = \text{const}$, are described by the relation

$$x^2 + \left(y - \frac{1 + \kappa^2}{1 - \kappa^2} f\right) = \frac{4f^2 \kappa^2}{(1 - \kappa^2)^2} \qquad (5.42)$$

The geometric figures are circles whose radii are equal to

$$|r| = \left| \frac{2f\kappa}{1 - \kappa^2} \right|$$

and whose centers are displaced by

$$b = \frac{(1 + \kappa^2)f}{1 - \kappa^2}$$

along the y axis. Since the depth f is not usually known explicitly, we eliminate it and obtain

$$\left| \frac{r}{b} \right| = \frac{2\kappa}{1 + \kappa^2} \qquad \text{or} \qquad \kappa = \left| \frac{b}{r} \right| \pm \left(\frac{b^2}{r^2} - 1 \right)^{1/2} \qquad (5.43)$$

The reduced heat transfer rate $\dot{Q}_L$ can be calculated with the aid of Eq. (5.41). Thus

$$\dot{Q}_L = \frac{2\pi kT}{\ln \kappa} = \frac{2\pi kT}{\cosh^{-1} |b/r|} \qquad (5.44)$$

The parameter κ is given by Eq. (5.43) with the aid of the auxiliary equation from Sec. 5.3.1. This heat transfer is generated by a pipe or cable of radius r at a depth b in a medium of thermal conductivity k; its surface temperature exceeds that of the surface of the ground by an amount T. Equation (5.44) also shows that the reduced shape factor is

$$S_L = \frac{2\pi}{\ln \kappa} = \frac{2\pi}{\cosh^{-1} |b/r|} \qquad (5.45)$$

in conformity with Eq. (5.37). Thus we reestablish the validity of all consequences of this equation.

The method of fictitious sources and sinks can be applied to solve three-dimensional problems and often leads to an answer also in cases where no conformal mapping into a simple geometrical shape is known.

5.5 GRAPHICAL METHODS

The properties of the two orthogonal networks, the isotherms, and the heat-flow streamlines, can be exploited to determine shape factors by graphical means. Figure 5.7 depicts a portion of such a network with curves of $T = \text{const}$ and $\psi = \text{const}$. The heat transfer rate between streamlines ψ' and ψ'' from isotherm T' to isotherm T'' can be put in the form

$$\dot{Q}' = kL(\psi' - \psi'') = \frac{kL \, \Delta s(T' - T'')}{\Delta n} \qquad (5.46)$$

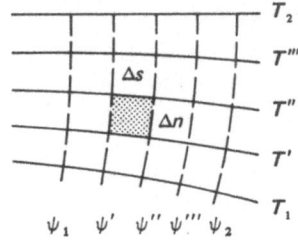

Figure 5.7 Isotherms $T = $ const and heat-flow stream-lines $\psi = $ const.

as seen from Eq. (5.23). Here, as usual, L denotes the thickness at right angles to the plane of the drawing. If we assign uniform temperature differences to sequential isotherms as well as equal differences to sequential streamlines in the whole network, we render the ratio $\Delta s/\Delta n$ constant, because Eq. (5.46) proves that

$$\frac{T' - T''}{\Delta n} = \frac{\psi' - \psi''}{\Delta s} \tag{5.47}$$

The reader may recall that the differential version of this equation was derived earlier as Eq. (5.22). For the sake of simplicity, it is convenient to choose equal magnitudes for both quantities, thus creating a mesh of squares.

The total heat transfer rate $\dot{Q}$, between streamlines ψ_1 and ψ_2 over a total of i_s thermal streamlines follows from Eq. (5.46), and is

$$\dot{Q} = kL(\psi_1 - \psi_2) = kLi_s \frac{\Delta s}{\Delta n} (T' - T'')$$

An equal heat transfer rate $\dot{Q}$ crosses from isotherm T'' to isotherm T''', and so on. We denote the number of successive isotherms between T_1 and T_2 by i_n and note that we then consider i_n such expressions for $\dot{Q}$. We now form a sum to eliminate the intermediate temperatures, and are led to the relation

$$i_n\dot{Q} = i_n kL(\psi_1 - \psi_2) = kLi_s \frac{\Delta s}{\Delta n} (T_1 - T_2)$$

This allows us to derive the following expression for the reduced shape factor:

$$S_L = \frac{\psi_1 - \psi_2}{T_1 - T_2} = \frac{i_s}{i_n} \frac{\Delta s}{\Delta n} \tag{5.48}$$

The application of this method is illustrated with the example of the insulation of a ship's hull, depicted in Fig. 5.8. The insulation is partly interrupted by a rib. The rib is assumed to be infinitely thin and its thermal conductivity is supposed to be infinitely large. It is best to begin the drawing of the isotherms and streamlines in a region where the elements of the mesh are still almost true squares. From there we extend the lines in both directions over the whole field, correcting reciprocally as we proceed. In the neighborhood of the rib the squares acquire curved sides. For checking purposes it is possible to draw a finer mesh in one curvilinear "square," as indicated in Fig. 5.8. The finer elements of the network are more nearly true squares. Exceptions occur near singular points, in this case near the top and bottom of the rib. Here we encounter "squares" of more than four sides which means that the principle of orthogonality no longer holds locally.

In a mesh element circumscribed by curved lines, the diagonals continue to be perpendicular to each other and the sums of opposite sides form a constant ratio. The graphical method, whose execution requires no further aids, is always convergent and its accuracy increases as smaller meshes are chosen. The accuracy of the drawing process places a natural upper bound to achievable overall accuracy. It is necessary to add here that the imposition of a fixed side ratio (conveniently: $\Delta s/\Delta n = 1$) makes it possible to choose an integer for at most one of the numbers i_s, i_n, because the shape factor S_L has an intrinsic value which is independent of the details of the drawing.

Figure 5.8 and Eq. (5.48) yield the numerical value

$$S_L = \frac{i_s}{i_n} \frac{\Delta s}{\Delta n} = \frac{7.3}{6} \times 1 = 1.22$$

Without the rib we had $S_L = 6/6 = 1.0$. This means that the provision of a rib increases the heat flow rate by a factor of $1.22/1.0 = 1.22$.

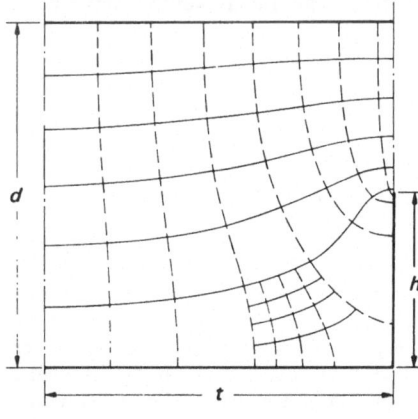

Figure 5.8 Insulation of a ship's hull with reinforcing rib: d, thickness of insulation; h, height of rib; t, half of the rib pitch ($t/d = 1$; $h/d = 0.5$).

5.6 RELAXATION METHOD

In the preceding section we described the graphical method whose essence it was to determine the corresponding number of heat-flow streamlines to match a given number of isotherms as well as to determine their geometric shape. The essence of the relaxation method, which we are about to expound, is numerically to determine the temperature field at preselected points.

The heat-conduction process is modeled with the aid of a set of rods forming a network created by a mesh of size Δx times Δy, Fig. 5.9. For simplicity, we choose $\Delta x = \Delta y$. The heat transfer rate from point 1 to point 0 is

$$\dot{Q}_{10} = \frac{k \, \Delta y L (T_1 - T_0)}{\Delta x} = kL(T_1 - T_0) \tag{5.49}$$

Similarly, $\dot{Q}_{20} = kL(T_2 - T_0)$, and so on. The sum of all heat transfer rates from points 1 to 4 with respect to 0 can be interpreted as the capacity of a heat sink located at point 0. In steady-state this capacity must be equal to zero. We introduce the reduced heat transfer rate $\dot{Q}' = \dot{Q}/(kL)$, where L denotes the thickness perpendicular to the plane of the drawing—as in all two-dimensional problems. We can then write down a heat balance for point 0 of the form

$$\dot{Q}' = T_1 + T_2 + T_3 + T_4 - 4T_0 \tag{5.50}$$

In steady-state, with $\dot{Q} = 0$, this becomes

$$T_0 = \tfrac{1}{4}(T_1 + T_2 + T_3 + T_4) \tag{5.51}$$

The preceding equation is an expression of the mean value theorem of

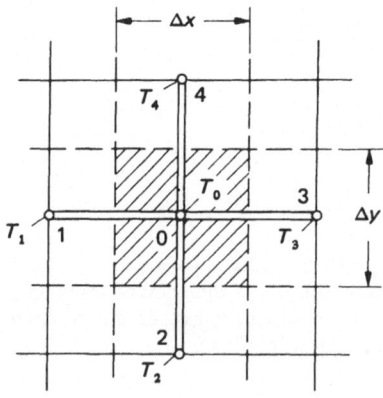

Figure 5.9 Illustration of the relaxation method. The homogeneous body is modeled with the aid of a network of heat-conducting rods.

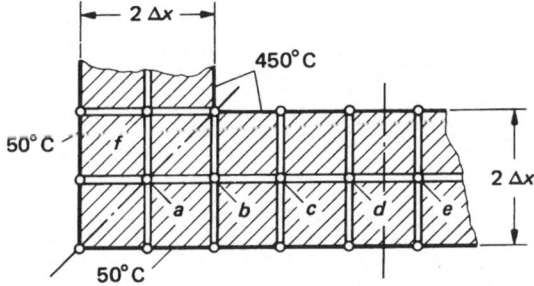

Figure 5.10 Illustrating the application of the relaxation method for a corner of a furnace with constant temperatures on the external surfaces.

potential theory, according to which the value of the potential at a point is equal to the arithmetic average of its values on a boundary surrounding this point.

The relaxation method consists in the following: we subdivide the body into elements. This is done intuitively, and on the basis of our optimal understanding of that body's behavior in the process we assign, by way of first approximation, guessed values of temperature at the nodal points. We then verify, point by point, whether the relation

$$\sum_1^4 (T_i - 4T_0) = 0$$

is satisfied. Remaining non-zero residues are eliminated in an order corresponding to their magnitude. The new residues in the adjoining points are again eliminated, until all numbers in the network attain constant values. The system is then said to have "relaxed." The accuracy of the procedure increases as the network is made finer, a natural limit being reached through the choice of the number of significant digits in each value.

Example 5.4 Figure 5.10 depicts the corner of a furnace whose external dimensions are $9 \, \Delta x \times 9 \, \Delta x$ and whose wall thickness is $2 \, \Delta x$. We are to determine the temperatures at the nodal points a to f, where the pairs b, f, as well as d, e are located symmetrically. The internal walls are kept at 450°C, whereas the external walls are at 50°C. Table 5.2 illustrates the calculation. The reader should note that the values of $\dot{Q}'$ and T are expressed in degrees Celsius, but the sign °C has been left out for simplicity. In Step 1 we have assigned the mean value between inner and outer temperature $[\frac{1}{2}(450 + 50) = 250]$ to points a

Table 5.2 Relaxation method applied to the calculation of the influence of corners on the heat loss of a furnace (see Fig. 5.10)

Step no.	f		a		b		c		d		e	
	$\dot{Q}'$	T	$\dot{Q}'$	T	$\dot{Q}'$	T	$\dot{Q}'$	T	$\dot{Q}'$	T	$\dot{Q}'$	T
1	0	250	-400	250	0	250	0	250	0	250	0	250
2	-100	250	0	150	-100	250	0	250	0	250	0	250
3	0	225	-50	150	0	225	-25	250	0	250	0	250
4	-13	225	2	137	-13	225	-25	250	-6	250	-6	250
5	-19	225	2	137	-19	225	-1	244	-6	250	-6	250
6	1	220	-8	137	1	220	-6	244	-6	250	-6	250
7	-1	220	0	135	-1	220	-6	244	-6	250	-6	250
8	-3	220	0	135	-3	220	2	242	-8	250	-8	250
9	-3	220	0	135	-3	220	0	242	-2	248	-2	248
10	1	219	-2	135	1	219	-1	242	-2	248	-2	248

through f. This yields $\dot{Q}' = 0$ everywhere, except for point a where the residual is $Q'_a = -400$. In Step 2 we reduce T_a by $400/4 = 100$ to a value 150. This generates the residuals $Q'_b = Q'_f = -100$ which are eliminated in Step 3 by reducing T_b and T_f by $100/4 = 25$ to a value of 225. After 10 steps the numbers do not change. The heat transfer rate from the inner wall to points b, c, and d is calculated with the aid of Eq. (5.49). We obtain

$$\dot{Q}_I = kL(231 + 208 + 202)K = kL \times 641 \text{ K}$$

From points a, b, c, and d to the outside we calculate

$$\dot{Q}_{II} = kL(85 + 169 + 192 + 198)K = kL \times 644 \text{ K}$$

On the average, the heat transfer rate through the eight partial surfaces of the exterior of the furnace is

$$\dot{Q}_m = 8 \times kL \times 642.5 \text{ K} = kL \times 5140 \text{ K}$$

We now consider the interior surface alone and disregard the effect of the corners. The heat transfer rate from it to the outside would be

$$\dot{Q}_i = 4 \times 5 \, \Delta x \times kL \times \frac{400 \text{ K}}{2 \, \Delta x} = kL \times 4000 \text{ K}$$

From the preceding calculation, we conclude that the presence of the corners has increased the heat loss by a factor $\dot{Q}_m / \dot{Q}_i = 1.29$.

5.7 EXPERIMENTAL ANALOGS

Laplace's differential equation for a potential describes a variety of physical phenomena. It is, therefore, possible to attempt to solve problems in heat conduction with the aid of analog systems which are experimentally simpler to arrange. Only two methods are of practical importance.

5.7.1 Electrolytic Tank

In this method, the stream of heat flow is modeled with the aid of a steady-state electric current which is made to flow between two or more electrodes. The potential difference becomes analogous to the temperature difference. If the electrodes of the model are made similar on some (practically) arbitrary scale to the isotherms of the real system, it becomes possible to deduce the shape factor directly from Eq. (5.7).

In the case of an electric current, the ratio of the resistance of the conductor to the insulator is many orders of magnitude (powers of 10) larger than is the case with a heat current. For this reason it is possible to model isothermal and adiabatic surfaces simply and accurately. Very frequently, a dilute aqueous solution is employed as the electrolyte. Such a solution is ideal for the representation of a homogeneous and isotropic material. In order to avoid problems with corrosion, it is also possible to use paper coated with graphite. In this case it is necessary to be cautious, because often the electric conductivity of such sheets depends on direction; that is, the sheets are anisotropic.

5.7.2 Soap-film Analogy

According to an equation derived by Laplace, the pressure difference, Δp, sustained by a curved liquid surface is

$$\Delta p = \sigma \left(\frac{1}{R_1} + \frac{1}{R_2} \right) \tag{5.52}$$

In this equation, σ denotes surface tension (a meaning used only in this section) and R_1 and R_2 are two radii of curvature in two arbitrarily placed, but mutually perpendicular, normal sectional planes. In the case of a soap bubble that develops two surfaces, it becomes necessary to multiply the right-hand side of Eq. (5.52) by a factor of 2. We denote the displacement of a point on a surface above its reference level by $z(x, y)$ and recall that the radius of curvature is then equal to

$$R = \frac{(1 + z')^{3/2}}{z''} \tag{5.53}$$

where $z' = \partial z / \partial x$ or $\partial z / \partial y$ if the x and y directions are chosen for Eq. (5.52). If we arrange conditions to satisfy the inequality $z' \ll 1$, such as, for example, $z' \ll 1/100$, it becomes possible to neglect it in Eq. (5.53) and to resort to the approximation

$$\frac{\Delta p}{2\sigma} = \frac{1}{R_1} + \frac{1}{R_2} = \frac{\partial^2 z}{\partial x^2} + \frac{\partial^2 z}{\partial y^2} \tag{5.54}$$

This is Poisson's differential equation which is analogous to the heat conduction equation in the presence of sources, that is to

$$\frac{\dot{q}'}{k} = \frac{\partial^2 T}{\partial x^2} + \frac{\partial^2 T}{\partial y^2} \tag{5.55}$$

As usual, $\dot{q}'$ is the volumetric source strength. If we put $\Delta p = 0$, or analogously, $\dot{q}' = 0$, both equations simplify to the Laplace equation.

Figure 5.11 illustrates the application of the method to the case of two eccentric circular pipes. The displacement $z(x, y)$ of the soap bubble

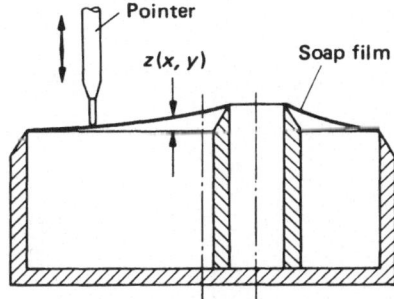

Figure 5.11 Soap-film analogy illustrated with the aid of the example of two eccentric cylinders.

above the contours of the pipes, each of which has been provided with a knife-edge circumference, is a measure of the temperature $T(x, y)$. The displacement $z(x, y)$ can be measured with the aid of a pointer and a dial gauge. Such measurements lead to results which are surprisingly close to analytic calculations. The soap film can also be used as a curved mirror in that one observes the deformation in a beam of parallel light. In this way it is possible to measure directly the gradients at the edges, employing Eq. (5.5) to determine the shape factor of the arrangement under study.

5.8 SUMMARY OF SHAPE FACTORS

Reference [5.4] contains an extensive list of values of the shape factor S which was defined in Sec. 5.1 with the aid of Eq. (5.4). The listing covers a range of generalized plate problems. We reproduce here a smaller set, which represents especially interesting cases from the practical point of view. The basic shape is defined in Figs. 5.12a and b; it consists of a strip of length $2t$, width d, and thickness L, measured at right angles to the plane of the drawing. Perpendicular to the isothermal edge at the bottom— maintained at temperature T_1—we have added very thin splitter plates or ribs of height h separated by a distance $2t$; these are also maintained at T_1.

In the limiting case when $h = 0$ we recover the simple-plate problem and we may write, on the basis of Eqs. (3.7) and (5.4), that

$$S_{L,\text{plate}} = \frac{S_{\text{plate}}}{L} = \frac{2t}{d} \tag{5.56}$$

In the general case when the rib height is finite, the two variants depicted in Figs. 5.12a and b lead to the same shape factor. That factor was evaluated in [5.4] with the aid of elliptic functions and represented in the form

$$S_{La} = S_{Lb} = S_{L,\text{plate}} + 2S_{LS} = \frac{2t}{d} + 2S_{LS} \tag{5.57}$$

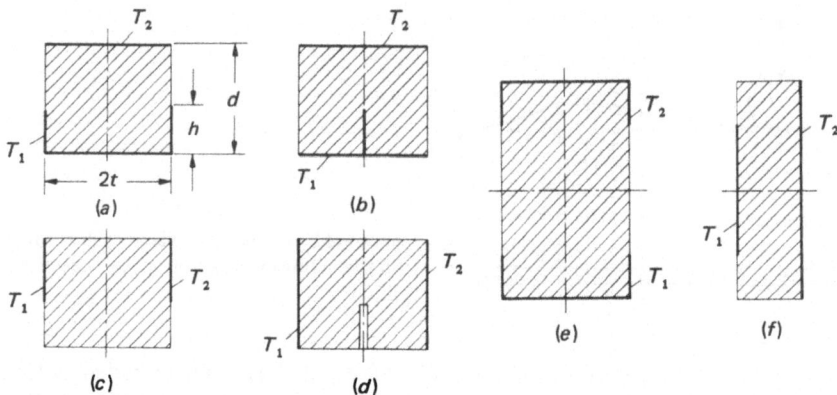

Figure 5.12a–f Heat conduction in plane slabs provided with splitter plates or ribs. Edges drawn with thin lines are adiabatic.

The subscripts a and b correspond to Figs. 5.12a and b, and S_{LS} represents the additional heat loss, or perturbation, introduced by the ribs which can be seen plotted in Fig. 5.13 in terms of the geometrical parameters $2t/d$ and h/d.

Reference [5.4] quotes the analytic expression

$$S_{LS\infty} = \frac{1}{\pi} \ln \frac{2}{1 - \sin\left[(\pi/2)(2h/d - 1)\right]} \tag{5.58}$$

which is valid in the limiting case when $2t/d \to \infty$. Figure 5.13 shows that this asymptotic value prevails from $2t/d \geqslant 2$ and constitutes a very good approximation for S_{LS}.

It was shown in Sec. 5.3 that in two-dimensional, potential fields it is possible to interchange the parts played by streamlines and potential lines. When this is done, adiabatic lines transform into isotherms and vice versa. Simultaneously, the shape factor becomes equal to the inverse of the one in the original arrangement. The configurations depicted in Figs. 5.12c and d are such inverses of the ones shown in Fig. 5.12a and b. It follows that their shape factors become

$$S_{Lc} = S_{Ld} = \frac{1}{S_{La}} = \frac{1}{S_{Lb}} \tag{5.59}$$

Once again, the subscripts a, b, c, and d relate to the corresponding diagram in Figs. 5.12a–d.

It was shown in Sec. 5.1 that form factors must be added when heat flows are parallel, and their inverses must be added when arranged in series. The arrangement shown in Fig. 5.12e is symmetric with respect to a

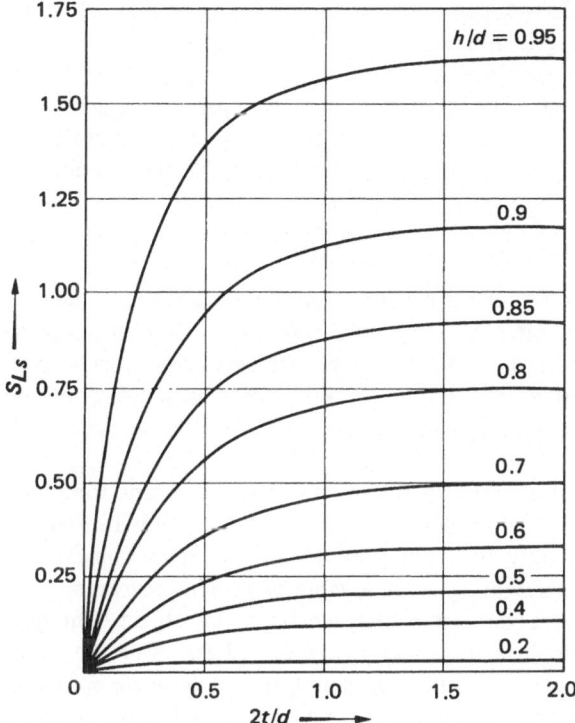

Figure 5.13 Perturbation shape factor S_{LS} as a function of the geometrical parameters $2t/d$ and h/d.

horizontal axis of symmetry and can, therefore, be regarded as two arrangements of Fig. 5.12a connected in series. Employing Eq. (5.57) we then find that

$$S_{Le} = \frac{1}{1/S_{La} + 1/S_{La}} = \frac{1}{2}\frac{2t}{d} + S_{LS} \tag{5.60}$$

The arrangement from Fig. 5.12a is itself symmetric with respect to a vertical axis and can be regarded as two strips of length t connected in parallel; its shape factor (written without a subscript) is, thus,

$$S_L = \frac{1}{2}S_{La} = \frac{1}{2}\frac{2t}{d} + S_{LS} = S_{Le} \tag{5.61}$$

Such an arrangement corresponds exactly to the case of the insulation of a ship's hull with a rib, analyzed earlier in Sec. 5.5. Referring to the diagram of Fig. 5.13 and employing the original values of the parameters, namely

$2t/d = 2.0$ and $h/d = 0.5$, we can retrieve the value $S_L = 1.22$ obtained earlier by an entirely different route.

In turn, we imagine that the arrangement in Fig. 5.12e has been cut along the vertical line of symmetry forming two parallel regions and interchange the plates at the end surfaces. In this manner we recreate the configuration of Fig. 5.12f and calculate its shape factor by the relation

$$S_{Lf} = \frac{1}{(1/2)S_{Le}} = \frac{2}{[(1/2)/(2t/d)] + S_{LS}} \tag{5.62}$$

The next class of two-dimensional heat conduction problems that we wish to consider involves circular boundaries; that is, we now propose to analyze generalized cylindrical problems. Such problems have already been discussed in another context in Sec. 5.3.1. For this reason, we continue with the last important grouping, namely with single bodies surrounded by an infinite medium.

Let us now once more examine the simple basic solutions for plate, cylinder, and sphere, Eqs. (3.7)–(3.9). It is important to note here that the case of the sphere is the only one that leads to a finite heat transfer rate upon letting $r_e \to \infty$ while retaining a finite value of r_i. This corresponds to an external wall placed at a very large distance from the object. The heat transfer rate for this spherical case is given by Eq. (3.10). For the shape factor of a very thick spherical shell (considered also as a single sphere in an infinite medium) of inner radius $r_i = r_1 = r$ on which the temperature T_1 is maintained with the external boundary $r_e = r_2 \to \infty$ and T_2, we find

$$S = 4\pi r \tag{5.63}$$

Table 5.3 lists shape factors for a variety of arrangements involving an isothermal sphere or hemisphere and an isothermal very thin circular or semicircular disk, each surrounded by an infinite or semi-infinite medium. Reference [5.5] contains an extensive listing of form factors for generalized plate, cylinder, and single-body arrangements from which Table 5.3 has been abstracted.

Example 5.5 A cylindrical hot-water reservoir (with vertical center line) is placed near a central thermal plant. The diameter of the reservoir is $d = 5$ m. The problem is to give a rough estimate of the heat loss from the circular bottom of the reservoir of temperature $T_1 = 50°C$ to the ground ($k = 0.6$ W/m K) whose temperature is $T_2 = 10°C$. Assuming that no heat is transferred between the air and the surface of the ground, we look up Table 5.3 and calculate

$$\dot{Q} = kS(T_1 - T_2) = k \times 2d \times (T_1 - T_2) = 240 \text{ W}$$

Table 5.3 Shape factors for spheres or disks in a medium of large extent (broken lines delineate adiabatic boundaries) [5.5]

Explanation of configurations (valid for sphere or disk)	Shape factor Sphere Disk
Medium boundless on all sides T_2 finite at very large distance Disk: $d/r = 0$	(1) Sphere $S = 4\pi r$ (2) Disk $S = 8r$
Medium semi-infinite Bounding plane partly adiabatic T_2 finite at very large distance	(3) Hemisphere $S = 2\pi r$ (4) Disk—one side $S = 4r$
Medium semi-infinite T_2 finite at very large distance and on bounding plane Equations valid only for $t/r \geqslant 2.0$	(5) Sphere $$S = \frac{4\pi r}{1 - r/2t}$$ (6) Disk $$S = \frac{4\pi r}{\frac{1}{2}\pi - \tan^{-1}(r/2t)}$$
Medium semi-infinite Bounding plane adiabatic T_2 finite at very large distance Equations valid only for $t/2r \geqslant 2.0$	(7) Sphere $$S = \frac{4\pi r}{1 + r/2t}$$ (8) Disk $$S = \frac{4\pi r}{\frac{1}{2}\pi + \tan^{-1}(r/2t)}$$

The shape factor $S = 4r = 2d$ employed for the estimate corresponds to the case of one side of a disk touching a semi-infinite medium, the second case in Table 5.3.

SIX

NONSTEADY ONE-DIMENSIONAL HEAT CONDUCTION

6.1 FUNDAMENTAL SOLUTIONS OF THE FOURIER EQUATION

The present chapter is devoted to a study of several solutions and methods of solving the one-dimensional Fourier equation

$$\frac{\partial T}{\partial t} = a \frac{\partial^2 T}{\partial x^2} \tag{6.1}$$

This is a linear, homogeneous partial differential equation of second order with constant coefficients, if it is permissible to assume that the thermal diffusivity $a = k/\rho c_p$ is constant. Dimensional considerations, to be repeated in greater detail in Sec. 6.2, allow us to guess that there must exist solutions of the form

$$T(x, t) = \Theta\left(\frac{x}{2(at)^{1/2}}\right) = \Theta(\xi) \tag{6.2}$$

In this solution T does not depend separately on x and t, but on the composite variable $\xi = x/2(at)^{1/2}$. Substitution of Eq. (6.2) into Eq. (6.1) leads to the ordinary differential equation

$$\Theta'' + 2\xi\Theta' = 0 \tag{6.3}$$

which possesses the general solution

$$\Theta = C_1 \int_0^{x/2(at)^{1/2}} \exp(-\xi^2)\,d\xi + C_2 \qquad (6.4)$$

The function

$$\operatorname{erf} z = \frac{2}{\pi^{1/2}} \int_0^z \exp(-\xi^2)\,d\xi \qquad (6.5)$$

represents Gauss's error integral, called error function for short. This function has the following properties:

$$\operatorname{erf} 0 = 0 \qquad \operatorname{erf} \infty = 1 \qquad \operatorname{erf}(-z) = -\operatorname{erf} z$$

By a suitable choice of the constants C_1 and C_2, we obtain the special solution

$$T = \frac{2T_c}{\pi^{1/2}} \int_0^{x/2(at)^{1/2}} \exp(-\xi^2)\,d\xi = T_c \operatorname{erf}\left(\frac{x}{2(at)^{1/2}}\right) \qquad (6.6)$$

The preceding solution of Eq. (6.1) satisfies the following boundary and initial conditions:

$$T = T_c \qquad \text{for all } x > 0 \text{ and } t = 0$$
$$T = -T_c \qquad \text{for all } x < 0 \text{ and } t = 0$$
$$T = 0 \qquad \text{for all } t > 0 \text{ and } x = 0$$

Since the Fourier equation (6.1) is linear, we conclude that derivatives (i.e., results of linear operations) of solutions with respect to x or t are also its solutions. Differentiating Eq. (6.4) with respect to x, we obtain

$$T = \frac{C}{2(at)^{1/2}} \exp(-\xi^2) \qquad (6.7)$$

and upon repeated differentiation

$$T = -\frac{Cx}{4(at)^{3/2}} \exp(-\xi^2) \qquad (6.8)$$

The last two solutions of Eq. (6.1) are called its fundamental solutions. Their application will be discussed in Chapter 7.

Additional solutions of the Fourier equation which are important in practical applications can be obtained by introducing Daniel Bernoulli's product assumption. Instead of studying Eq. (6.1) we discuss its more general form

$$\frac{\partial T}{\partial t} = a\left(\frac{\partial^2 T}{\partial r^2} + \frac{n}{r}\frac{\partial T}{\partial r}\right) \tag{6.9}$$

with $n = 0$ for plates, $n = 1$ for cylinders, and $n = 2$ for spheres. The product assumption takes the form

$$T = \phi(t) \times \psi(r) \tag{6.10}$$

where ϕ is a function of t only, and ψ depends exclusively on r. Substitution of Eq. (6.10) into Eq. (6.9) leads us to the consideration of two ordinary differential equations

$$\frac{1}{a}\frac{\phi'}{\phi} = \pm\beta^2 = \frac{1}{\psi}\left(\psi'' + \frac{n\psi'}{r}\right) \tag{6.11}$$

The separated differential expressions must be equal to the same constant, denoted here by $\pm\beta^2$, because ϕ is independent of r and ψ is independent of t. We examine the solution of the left equation:

$$\phi = C \exp\left(\pm\beta^2 \, at\right) \tag{6.12}$$

and conclude that only the negative sign has a reasonable physical meaning in our context, because internal temperatures can only decrease with time in the absence of heat sources. The right-hand equation is solved by

$$\psi = C' \cos\beta r \qquad \text{and} \qquad \psi = C'' \sin\beta r \qquad \text{for } n = 0$$

$$\psi = C' J_0(\beta r) \qquad \text{and} \qquad \psi = C'' Y_0(\beta r) \qquad \text{for } n = 1$$

$$\psi = C' \frac{\sin\beta r}{\beta r} \qquad \text{and} \qquad \psi = C'' \frac{\cos\beta r}{\beta r} \qquad \text{for } n = 2$$

The solution of Eq. (6.9) can be recorded in the general form

$$T = C \exp\left(-\beta^2 \, at\right) \times \psi(\beta r) \tag{6.13}$$

The free constants C and β must be determined with reference to the initial and boundary conditions. The symbols J_0 and Y_0 denote Bessel functions of the first and second kind, respectively, both of zero order. We shall find applications for these solutions in Sec. 6.6.

6.2 DIMENSIONAL ANALYSIS

Since all terms of a properly formulated equation in physics—a quantity equation—are dimensionally homogeneous, it must be possible to cast it in dimensionless form. As far as differential equations as well as their initial and boundary conditions are concerned, the preceding observation applies to the variables as well as the coefficients. By this procedure we can identify the dimensionless arguments of the solution without finding the solution itself explicitly. In most cases, the number of such dimensionless arguments is smaller than the number of dimensional quantities which enter the problem. We now apply these remarks to the Fourier equation

$$\frac{\partial T}{\partial t} = a \frac{\partial^2 T}{\partial x^2} \tag{6.14}$$

We put $\Theta = T/T_c$ and $\xi = x/X$ thus defining a dimensionless temperature Θ and a dimensionless length variable ξ. Here T_c denotes a characteristic constant temperature and X a characteristic length. It is always very useful to choose T_c and X in such a way as to render initial or boundary conditions also dimensionless. For example, if T_c is chosen from among the boundary conditions, the latter will become a pure number—unity in this case. Similarly, it is useful to make X equal to half the plate thickness. In cases when there exists no characteristic time in the problem, it is useful to introduce a dimensionless time $\tau = at/X^2$. For example, this occurs when a temperature at a boundary increases or decreases step-wise. If the boundary is specified in the form of a time-dependent temperature $T_0(t)$, or if a time-varying heat flux $\dot{q}(t)$ enters the problem, it may not be possible to specify a single such dimensionless variable. With the preceding definitions, the Fourier equation acquires the dimensionless form

$$\frac{\partial \Theta}{\partial \tau} = \frac{\partial^2 \Theta}{\partial \xi^2} \tag{6.15}$$

whose solution is

$$\Theta = f(\xi, \tau) \qquad \text{or} \qquad \frac{T}{T_c} = f\left(\frac{x}{X}, \frac{at}{X^2}\right)$$

If the problem contains no characteristic length in its geometry or its boundaries, the solution must depend on a homogeneous product

$$\left(\frac{x}{X}\right)^a \frac{at}{X^2} \qquad (\alpha, \text{ a constant})$$

which is independent of X. It can be verified that the proper form is

$$\left(\frac{at}{x^2}\right)^\gamma \qquad (\gamma, \text{ a constant})$$

in which x has taken the place of X. In such cases the solution would become

$$\frac{T}{T_c} = f\left(\frac{at}{x^2}\right)$$

or, equivalently,

$$\frac{T}{T_c} = f\left(\frac{x}{2(at)^{1/2}}\right)$$

as we already found in Eq. (6.6). In problems involving a boundary condition of the third kind with respect to a wall perpendicular to x, with the positive direction of x pointing inward, we would have

$$-\left(\frac{\partial T}{\partial x}\right)_w = -\frac{T_w}{k/h} \tag{6.16}$$

assuming that the temperature of the fluid is $T_\infty = 0$. Employing the preceding dimensionless variables, we transform Eq. (6.16) to the form

$$\frac{1}{\Theta_w}\left(\frac{\partial \Theta}{\partial \xi}\right)_w = \frac{hX}{k} \tag{6.17}$$

Buckingham's Pi theorem applied to conduction problems with appropriate boundary conditions would show that the original seven physical quantities reduce by four when we introduce dimensionless quantities, because the problem under consideration is characterized by four fundamental dimensions: length, time, temperature, and heat. Mass plays no part in our problem.

In the sciences of fluid mechanics and heat transfer it has become accepted to give names to dimensionless characteristic numbers which honor important contributors to the subject. In this case we use the nomenclature:

Fourier number $\qquad$ $\mathrm{Fo} = \dfrac{at}{X^2} \qquad$ or $\qquad \dfrac{at}{x^2}$

Biot number $\qquad$ $\mathrm{Bi} = \dfrac{hX}{k} \qquad$ or $\qquad \dfrac{hx}{k}$

In this notation, the solution of the Fourier equation for one-dimensional nonsteady heat conduction, with a single boundary condition of the third kind and no characteristic time in the problem, assumes the form

$$\frac{T}{T_c} = f\left(\frac{x}{X}, \mathrm{Fo}, \mathrm{Bi}\right) \tag{6.18}$$

6.3 SEMI-INFINITE BODY

Equation (6.6), derived in Sec. 6.1, namely,

$$\frac{T}{T_c} = \frac{2}{\pi^{1/2}} \int_0^{x/[2(at)^{1/2}]} \exp\left(-\xi^2\right) d\xi = \text{erf } \xi \tag{6.19}$$

with the abbreviation $\xi = x/[2(at)^{1/2}]$, describes the temperature distribution in a semi-infinite body of uniform temperature T_c whose external surface temperature has been suddenly cooled to $T_w = 0$ at instant $t = 0$ and then maintained at this level. The rate of cooling $-\partial T/\partial t$ is equal to

$$-\frac{\partial T}{\partial t} = \frac{T_c x}{2(\pi a t^3)^{1/2}} \exp\left(-\xi^2\right) \tag{6.20}$$

and the heat flux is

$$\dot{q}_x = -k\frac{\partial T}{\partial x} = -\frac{kT_c}{(\pi at)^{1/2}} \exp\left(-\xi^2\right) = -\frac{bT_c}{(\pi t)^{1/2}} \exp\left(-\xi^2\right) \tag{6.21}$$

Here

$$b = \frac{k}{a^{1/2}} = (k\rho c_p)^{1/2}$$

is the coefficient of heat penetration known to the student from Sec. 2.4 where it first appeared. At the surface ($\xi = 0$; $t > 0$) the heat flux is

$$\dot{q}_w = -\frac{bT_c}{(\pi t)^{1/2}}$$

The quantity of heat Q_{01} which has crossed surface area A from instant $t = 0$ to $t = t_1$ is obtained by integration as

$$Q_{01} = A \int_0^{t_1} \dot{q}_w \, dt = -\frac{2}{\pi^{1/2}} AbT_c t_1^{1/2} \tag{6.22}$$

The sudden cooling of the surface of a semi-infinite body to a constant temperature T_w occurs, for example, under a fresh snowfall when $T_w = 0°C$. The amount of snow which melts at the surface over area A depends in the first instance on the quantity of heat Q_{01} which crossed; according to Eq. (6.22), it is proportional to the coefficient of heat penetration b of the ground. For this reason, snow melts fastest on a metallic surface, and progressively more slowly on a massive rock, porous soil, timber, and a grass patch (cf. the data of Table 2.6).

If the initial temperature of the semi-infinite body is $T = 0$ and the temperature of its surface is suddenly raised to T_c, Eq. (6.19) must be replaced by

$$\frac{T}{T_c} = 1 - \text{erf } \xi = \text{erfc } \xi = \frac{2}{\pi^{1/2}} \int_{x/[2(at)^{1/2}]}^{\infty} \exp\left(-\xi^2\right) d\xi \quad (6.23)$$

The function erfc $\xi = 1 - \text{erf } \xi$ is called the complementary error function; it assumes the values erfc $0 = 1$, erfc $\infty = 0$, erfc $(-z) = 2 - \text{erfc } z$. Equations (6.20)–(6.22) remain valid, except for a change in sign. Values of the error function and related functions are given in Table G.1 of the Appendix; they are seen graphed in Fig. 6.1.

Example 6.1 At the midpoint $T/T_c = 0.5$, Eq. (6.19) gives $\xi = 0.477$ or $at/x^2 = 1.099$. Table 6.1 lists the times at which this midpoint temperature $T = 0.5 \, T_c$ is attained in different materials at depths of $x = 0.01$ m, 0.1 m, and 1.0 m.

Solutions of the Fourier equation

$$\frac{\partial T}{\partial t} = a \frac{\partial^2 T}{\partial x^2}$$

can be utilized for the calculation of heat fluxes. We differentiate both sides of the equation with respect to x to obtain

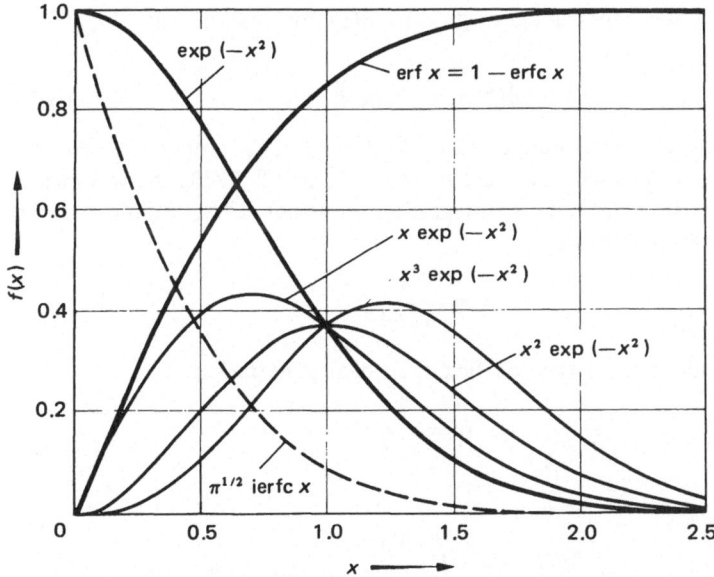

Figure 6.1 Error function erf x and related functions.

Table 6.1 Times at which midpoint temperature $T/T_c = 0.5$ is achieved in semi-infinite body

	Copper	Iron	Glass	Wood
a, 10^{-6} m^2/s	107	16.3	0.617	0.139
Depth $x = 0.01$ m	1.03 s	6.77 s	2.97 min	13.2 min
0.1 m	1.72 min	11.3 min	4.95 h	22.0 h
1.0 m	2.87 h	18.8 h	20.6 d	91.6 d

$$\frac{\partial}{\partial t}\left(\frac{\partial T}{\partial x}\right) = a\,\frac{\partial^2}{\partial x^2}\left(\frac{\partial T}{\partial x}\right) \tag{6.24}$$

and substitute $\dot{q} = -k\,\partial T/\partial x$. Hence

$$\frac{\partial \dot{q}}{\partial t} = a\,\frac{\partial^2 \dot{q}}{\partial x^2} \tag{6.25}$$

The heat flux $\dot{q}(x, t)$ satisfies the same differential equation (but not necessarily the same initial or boundary conditions) as the temperature $T(x, t)$, and all solutions obtained so far can be put to a new practical use on condition that due attention is paid to the initial and boundary conditions. Thus, when a constant heat flux $\dot{q}_w$ is suddenly applied to a semi-infinite body whose interior is free from temperature gradients $[\dot{q}(x, 0) = 0]$, we can calculate $\dot{q}(x, t) = \dot{q}(\xi)$ by analogy with Eq. (6.23) and write

$$\dot{q}(\xi) = \dot{q}_w \operatorname{erfc} \xi \tag{6.26}$$

Let us assume an initial temperature $T(x, 0) = 0$. The temperature distribution is now calculated by partial integration of $\partial T/\partial x = -\dot{q}/k$ which introduces a time-dependent integration constant $C(t)$. Employing Eq. (6.26) with the substitution

$$\xi = \frac{x}{2(at)^{1/2}}$$

and the introduction of the coefficient of heat penetration

$$b = \frac{k}{a^{1/2}}$$

we calculate

$$T(x, t) = -\frac{2(at)^{1/2}\dot{q}_w}{k} \int_0^{x/[2(at)^{1/2}]} \operatorname{erfc} \xi\, d\xi + C(t) \tag{6.27}$$

$$T(x, t) = -\frac{2\dot{q}_w t^{1/2}}{b} \left(\int_0^\infty \text{erfc } \xi \, d\xi - \int_{x/[2(at)^{1/2}]}^\infty \text{erfc } \xi \, d\xi \right) + C(t)$$

$$= -\frac{2\dot{q}_w t^{1/2}}{b} \left(\frac{1}{\pi^{1/2}} - i \text{ erfc } \xi \right) + C(t)$$

$$= \frac{2\dot{q}_w t^{1/2}}{b} \left[\frac{\exp(-\xi^2)}{\pi^{1/2}} - \xi \text{ erfc } \xi \right] \qquad \begin{array}{c} (6.27) \\ (\text{Cont.}) \end{array}$$

Equation (6.27) was derived with the use of the integral complementary error function

$$i \text{ erfc } \xi = \int_{x/[2(at)^{1/2}]}^\infty \text{erfc } \xi \, d\xi = \frac{\exp(-\xi^2)}{\pi^{1/2}} - \xi \text{ erfc } \xi$$

Values of this function are listed in Table G.1 of the Appendix and graphed in Fig. 6.1. At $x = 0$, we have ierfc $0 = 1/\pi^{1/2}$ and the initial condition $T(x, 0) = 0$ leads to

$$C(t) = \frac{2\dot{q}_w t^{1/2}}{\pi^{1/2} b}$$

The temperature variation at the surface is

$$T_w(t) = \frac{2\dot{q}_w t^{1/2}}{\pi^{1/2} b} \qquad (6.28)$$

Solutions for the semi-infinite body can be adapted to solve problems involving bodies of finite dimensions, as long as the thermal influences are confined to their surfaces. The above solutions constitute, therefore, usable approximations for short times.

6.4 TWO SEMI-INFINITE BODIES IN THERMAL CONTACT

We consider two semi-infinite bodies made of different materials, but sustaining equal, uniform temperature fields. We denote the two penetration depths by b_1 and b_2. A heat transfer rate $\dot{Q}$ is suddenly released at the contact surface. In these circumstances, Eq. (6.28) retains its validity. Since the bounding plane can store no heat and must have a uniform temperature $T_w(t)$, we observe that

$$\frac{\dot{Q}}{A} = \dot{q}_1 + \dot{q}_2$$

and according to Eq. (6.28), we may write that

$$T_w(t) = \frac{2}{\pi^{1/2}} \frac{\dot{q}_1}{b_1} t^{1/2} = \frac{2}{\pi^{1/2}} \frac{\dot{q}_2}{b_2} t^{1/2}$$

where $\dot{q}_1$ and $\dot{q}_2$ denote the time-independent heat fluxes in each of the respective regions. From this, we can derive the relations

$$\frac{\dot{q}_1}{\dot{q}_2} = \frac{b_1}{b_2} \quad \text{or} \quad \frac{\dot{q}_1}{\dot{Q}/A} = \frac{b_1}{b_1 + b_2} \quad \text{and} \quad T_w(t) = \frac{2\dot{Q}/A}{b_1 + b_2} \left(\frac{t}{\pi}\right)^{1/2}$$

$$(6.29)$$

Example 6.2 The changing temperature of the ground induced by radiation cooling during a cloudless, calm night can be described by Eq. (6.29) by way of approximation. We interpret $\dot{Q}/A$ as the difference between the heat flux gained and lost by the atmosphere through radiation, and assume for it the plausible value $\dot{Q}/A = -150$ W/m². We put $b_1 + b_2 = 1500$ W s$^{1/2}$/m² K, where b_2 for air can be neglected. Thus, for $t = 5$ h we calculate a temperature drop on the ground of $T_w \approx -15$ K. The preceding rough estimate offers a plausible explanation for the formation of frozen dew or hoarfrost on a night in late summer.

We now consider two bodies sustaining different, but uniform initial temperatures T_1 and T_2 whose thermal properties are denoted by k_1, a_1 and k_2, a_2, respectively. We let the two bodies be brought into sudden thermal contact at $t = 0$, Fig. 6.2.

After a very short time interval, the temperature at the contact surface will become equal on both sides, assuming a value denoted by T_m. Since

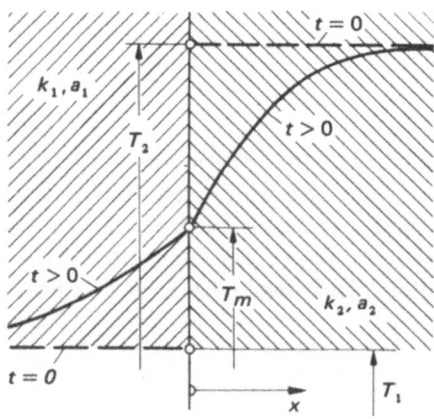

Figure 6.2 Contact temperature T_m between two semi-infinite bodies.

the boundary cannot store heat, we observe that for $t > 0$ we must have

$$T_m(x = -0) = T_m(x = +0) \quad \text{and} \quad k_1 \left(\frac{\partial T}{\partial x}\right)_{x=-0} = k_2 \left(\frac{\partial T}{\partial x}\right)_{x=+0}$$

In order to calculate the mean temperature T_m we assume, provisionally, that it remains independent of time. If we succeed in solving the system of equations with the aid of this stipulation, we shall conclude *post factum* that it was justified. If this is right, we can apply Eqs. (6.19) and (6.23) and calculate

for $x > 0$: $\qquad T(x, t) = (T_2 - T_m) \operatorname{erf} \xi_2$

for $x < 0$: $\qquad T(x, t) = (T_m - T_1) \operatorname{erfc} (-\xi_1)$

for $x > 0$: $\qquad \dfrac{\partial T}{\partial x} = \dfrac{(T_2 - T_m) \exp (-\xi_2^2)}{(\pi a_2 t)^{1/2}}$

for $x < 0$: $\qquad \dfrac{\partial T}{\partial x} = \dfrac{(T_m - T_1) \exp (-\xi_1^2)}{(\pi a_1 t)^{1/2}}$

where $\qquad \xi_1 = \dfrac{x}{2(a_1 t)^{1/2}} \quad \text{and} \quad \xi_2 = \dfrac{x}{2(a_2 t)^{1/2}}$

For the boundary $(x = 0)$, we obtain the condition

$$\frac{k_1(T_m - T_1)}{a_1^{1/2}} = \frac{k_2(T_2 - T_m)}{a_2^{1/2}}$$

or, using the abbreviations,

$$b_1 = \frac{k_1}{a_1^{1/2}} \quad \text{and} \quad b_2 = \frac{k_2}{a_2^{1/2}}$$

we get

$$\frac{T_m - T_1}{T_2 - T_m} = \frac{b_2}{b_1}$$

that is

$$\frac{T_m - T_1}{T_2 - T_1} = \frac{b_2}{b_1 + b_2} \tag{6.30}$$

We express the content of this equation in words: The contact temperature T_m, which is time-independent, is closer to the temperature of the body whose penetration depth is larger. This fact can be utilized for the measurement of b and, hence, also of k. This is done by measuring T_m after contact has been established with a body whose value of b is known. Equation (6.30) explains why different bodies of equal temperature do not give the same impression of being cold or warm when we touch them.

Example 6.3 The temperature of the skin of a human hand is $T_1 = 30°C$, its depth of penetration is estimated at $b_1 = 1000$ W $s^{1/2}/m^2$ K. On touching different bodies of the same temperature $T_2 = 100°C$, we would calculate the following contact temperatures:

Material	b_2, W $s^{1/2}/m^2$ K	T_m, °C
Copper	36,000	98
Iron	15,000	96
Soapstone	1,860	77
Wood	370	49
Foam rubber	40	33

The preceding calculation ignores the ability of a human body to regulate its temperature.

6.5 DUHAMEL'S THEOREM

The solutions to the Fourier equation derived in the preceding sections involved a single, sudden jump in the temperature at the boundary. These solutions can be comparatively easily adapted to apply to time-dependent boundary conditions by the use of a mathematical method first indicated by Duhamel (1833). We proceed to explain this principle referring the reader to Fig. 6.3.

The broken line describes the familiar case when the wall temperature T_w is made to jump at instant $t = 0$ from $T_w = 0$ to T_c and is kept at that level at all times $t > 0$. In the special case of a semi-infinite body with initial temperature $T(x, 0) = 0$, the temperature field is described by the equation

$$\frac{T(x, t)}{T_c} = \text{erfc } \xi$$

where

$$\xi = \frac{x}{2(at)^{1/2}}$$

with T_c chosen as the reference temperature. The relative temperature T/T_c retains a value of unity at $x = 0$ for $t > 0$ and we express this fact by saying colloquially that "The function $T(x, t)/T_c$ is the response of a system whose temperature is uniform at $t = 0$—in this particular case, of a

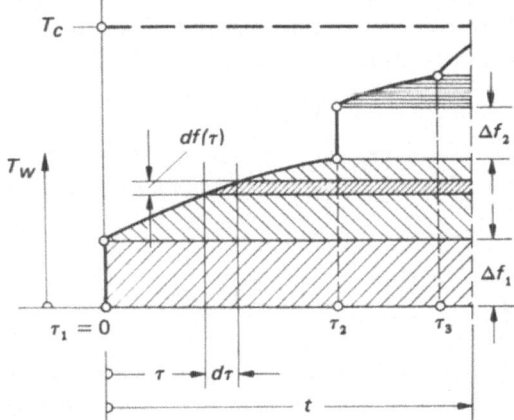

Figure 6.3 Illustrating the derivation of Duhamel's theorem.

semi-infinite body with $T(x, 0) = 0$—to an imposed disturbance consisting of a unit jump at $t = 0$." This clearly defined response to a unit jump will henceforth be denoted by $F(x, t)$.

The solid line in Fig. 6.3 represents an arbitrary variation $T_w(t)$ of wall temperature. In detail, we have assumed that in a time interval $0 \leqslant \tau \leqslant t$ there have occurred jump discontinuities at $\tau = \tau_1$ and $\tau = \tau_2$ and a kink at $\tau = \tau_3$. At instant $t = 0$, the present system, like the one before, is assumed to be characterized by a uniform temperature $T = 0$. For simplicity, we put $T_w(\tau) = f(\tau)$ and $\Delta T_w(\tau_i) = \Delta f_i$ at the jump discontinuities.

The basic idea of Duhamel's theorem rests on the application of the superposition principle which asserts that partial solutions of linear, ordinary or partial, differential equations may be added together. Realizing this idea, we note that the contribution of n jump discontinuities in the wall temperature, which occur in the interval $0 \leqslant \tau \leqslant t$, can be written down immediately in the form

$$T_{\text{jump}}(x, t) = \sum_{i=1}^{n} [\Delta f_i \times F(x, t - \tau_i)] \tag{6.31}$$

Equation (6.31) and all remaining formulas of this section continue to be valid for $t \leqslant \tau_n$ if each jump $F(x, t - \tau_i)$ is set equal to zero in the interval $t - \tau_i \leqslant 0$.

The jumps Δf_i of individual discontinuities serve as weighting factors for the responses $F(x, t - \tau_i)$, which have been normalized to a unit jump. However, their contribution must not be added at instant $\tau = 0$, but must be delayed to instant $\tau = \tau_i$. Thus the time when the influence is exerted is given by $t - \tau_i$. In regions where the temperature variation is continuous, we include first the differential contribution

$$dT(x, \tau) = df(\tau) \times F(x, t - \tau)$$

Extending formally by $d\tau$ and integrating over an interval $\tau_i \leqslant \tau \leqslant \tau_{i+1}$, which is bounded either by jump discontinuities, by kinks or by the instants $\tau = 0$ and $\tau = t$, we may write

$$T(x, \tau) \bigg|_i^{i+1} = \int_{\tau_i}^{\tau_{i+1}} \frac{df(\tau)}{d\tau} F[x, (t - \tau)] \, d\tau$$

The sum of all integrals over m such intervals in which the derivative $df(\tau)/d\tau$ remains continuous and nonvanishing may now be added to the contribution of the n jump discontinuities from Eq. (6.31). Hence, at instant t, we find that

$$T(x, t) = \sum_{i=1}^{n} [\Delta f_i \times F(x, t - \tau_i)] + \sum_{i=1}^{m} \left[\int_{\tau_i}^{\tau_{i+1}} \frac{df(\tau)}{d\tau} F(x, t - \tau) \, d\tau \right]$$

$$(6.32)$$

Kinks make no contribution to the solution; they operate merely as boundaries between two continuous intervals.

Figure 6.4 shows a number of strips—displaced successively to the right—each marked by a different style of hatching. These strips illustrate the way the temperature field is affected as time progresses. Equation (6.32) was derived from this point of view. This equation constitutes one form of Duhamel's theorem which is valid as long as our earlier assumptions remain valid. It is noteworthy that Duhamel's theorem can be applied to other (linear!) boundary conditions.

To illustrate our argument, we consider a semi-infinite body of initial temperature $T(x, 0) = 0$ whose wall temperature varies according to the following scheme:

$$T_w = T_1 \quad \text{for } 0 < t < \tau^* \quad \text{and} \quad T_w = T_2 \quad \text{for } t > \tau^*$$

Applying Eqs. (6.23) and (6.32) [or Eq. (6.31), because we have stipulated only two jumps], we write down the solution

$$T(x, t) = T_1 \, \text{erfc} \left[\frac{x}{2(at)^{1/2}} \right] + (T_2 - T_1) \, \text{erfc} \left[\frac{x}{2a^{1/2}(t - \tau^*)^{1/2}} \right]$$

$$(6.33)$$

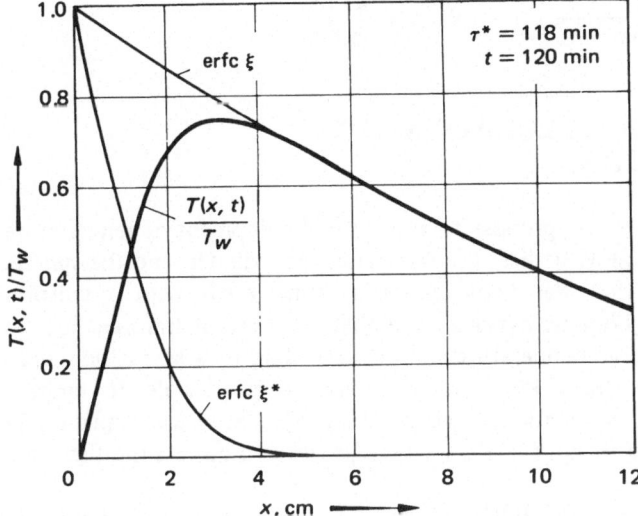

Figure 6.4 Heated wall of furnace with sudden chilling.

If at instant τ^* we impress temperature $T_2 = 0$ on the wall, we are led to the solution

$$T(x,\ t) = T_1 \left\{ \mathrm{erfc}\left[\frac{x}{2(at)^{1/2}} \right] - \mathrm{erfc}\left[\frac{x}{2a^{1/2}(t - \tau^*)^{1/2}} \right] \right\} \quad (6.34)$$

Example 6.4 The thick wall of an industrial furnace ($a = 10^{-6}\ \mathrm{m^2/s}$) and of uniform temperature $T(x, 0) = 0$ is heated suddenly so that its surface temperature jumps to T_w at $t = 0$. The initial temperature is again reduced to zero at $\tau^* = 118$ min. We wish to determine the temperature distribution at $t = 120$ min. We form the expressions

$$\xi = \frac{x}{2(at)^{1/2}} = \frac{x/m}{2(10^{-6} \times 120 \times 60)^{1/2}} = 5.89\ \frac{x}{m}$$

$$\xi^* = \frac{x}{2a^{1/2}(t - \tau^*)^{1/2}} = \frac{x/m}{2[10^{-6} \times (120 - 118) \times 60]^{1/2}} = 45.64\ \frac{x}{m}$$

The distribution $T(x, t)/T_w$ is thus

$$\frac{T(x,\ t)}{T_w} = \mathrm{erfc}\ \xi - \mathrm{erfc}\ \xi^*$$

as shown in Fig. 6.4. One curve arises from the other by stretching the values of the abscissas by the factor

$$\left(\frac{t}{t-\tau^*}\right)^{1/2} = 60^{1/2} = \frac{45.64}{5.89} = 7.75$$

6.6 TEMPERATURE EQUALIZATION IN SIMPLE BODIES

In the present section we propose to study the development of temperature fields in so-called simple bodies: the flat plate, the cylinder, and the sphere. We shall assume that the body exists at time $t = 0$ with a uniform temperature throughout its expanse, and that for $t > 0$ it transfers heat to a surrounding fluid at temperature $t_\infty = 0$ according to a set of boundary conditions of the third kind. In this case it is possible to employ Bernoulli's product assumption to good effect. The latter was explained in Sec. 6.1 and led us to Eq. (6.13). We now proceed to discuss the flat plate in greater detail.

We denote the plate thickness by $2X$ and the length coordinate measured from the center plane by x. With this notation we establish that

$$T = T_c \qquad \text{for } -X < x < X \qquad \text{and} \qquad t = 0$$

$$\frac{\partial T}{\partial x} = -\frac{h}{k}T \qquad \text{for} \quad x = X \qquad \text{and} \qquad t > 0$$

$$\frac{\partial T}{\partial x} = \frac{h}{k}T \qquad \text{for} \quad x = -X \qquad \text{and} \qquad t > 0$$

on the assumption that the temperature of the fluid has been set at zero. We stipulate that the coefficients a and h/k are constants. According to the form of the solution given in Eq. (6.13), namely

$$T = C \exp\left(-\beta^2 at\right) \times \psi(\beta x)$$

we may assume that the function of x alone is of the form

$$\psi = C' \cos \beta x + C'' \sin \beta x$$

Our assumptions assure that the temperature field must be symmetric around $x = 0$. Consequently, we include only the cosine function ($C'' = 0$) and put

$$T(x, t) = C \exp\left(-\beta^2 at\right) \times \cos \beta x \qquad (6.35)$$

where CC' has been replaced by C. The free constants C and β are now adjusted to suit the initial and boundary conditions. The boundary conditions at $\pm X$ lead to

$$\frac{\partial T}{\partial x} = C \exp\left(-\beta^2 at\right) \times (\mp\beta \sin \beta X) \qquad \text{for } x = \pm X$$

$$\frac{\partial T}{\partial x} = -\frac{h}{k} T = -\frac{h}{k} C \exp\left(-\beta^2 at\right) \times \cos \beta X \qquad \text{for } x = \mathsf{I} X$$

$$\frac{\partial T}{\partial x} = \frac{h}{k} T = \frac{h}{k} C \exp\left(-\beta^2 at\right) \times \cos \beta X \qquad \text{for } x = -X$$

These conditions lead to a transcendental equation for the constant β, which is of the form

$$\frac{h}{k}\cos \beta X = \beta \sin \beta X$$

Introducing the alternative constant $\delta = \beta X$ and the Biot number

$$\text{Bi} = \frac{hX}{k}$$

from Sec. 6.2, we are led to

$$\cot \delta = \frac{\delta}{\text{Bi}} \qquad \text{or} \qquad \delta \tan \delta = \text{Bi} \qquad (6.36)$$

Since the cotangent function is periodic, its intersection with the straight line δ/Bi yields an infinity of roots: $\delta_1, \delta_2, \ldots, \delta_m$, as we can infer from Fig. 6.5. These assume the following limiting values:

For $\text{Bi} = \infty$ (boundary condition of the first kind), we have $\delta_1 = \frac{1}{2}\pi$, $\delta_2 = \frac{3}{2}\pi$, $\delta_3 = \frac{5}{2}\pi, \ldots, \delta_m = (m - \frac{1}{2})\pi$.
For $\text{Bi} = 0$ (adiabatic wall), we have $\delta_1 = 0$, $\delta_2 = \pi$, $\delta_3 = 2\pi, \ldots,$ $\delta_m = (m-1)\pi$.

The numerical values of δ implied in Eq. (6.36) are called eigenvalues (or characteristic values) of the problem. The Fourier equation can be satisfied together with its boundary conditions only by the eigenvalues. The latter depend on the Biot number alone. The complete solution must be constructed in form of a sum of partial solutions and is

$$T(x, t) = \sum_{m=1}^{\infty}\left[C_m \exp\left(-\delta_m^2 \text{Fo}\right) \times \cos\frac{\delta_m x}{X}\right] \qquad (6.37)$$

where the Fourier number is defined as

$$\text{Fo} = \frac{at}{X^2}$$

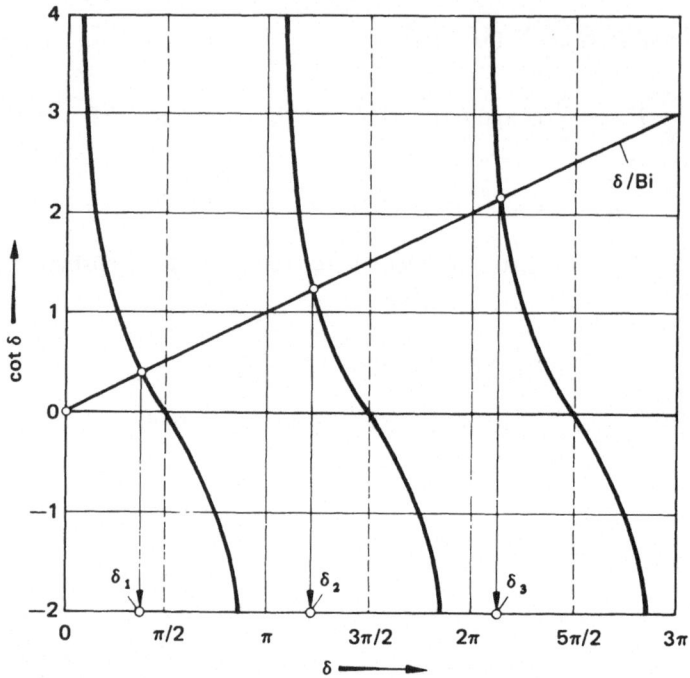

Figure 6.5 Illustrating the determination of the eigenvalues δ_m for a flat plate.

In order to satisfy the initial condition $T = T_c$ for $-X < x < X$, we must deal with the expansion of a given function in a trigonometric series, known as the Fourier series in honor of its discoverer. In the present case we encounter the additional peculiarity in that the factors δ_m must satisfy the transcendental Eq. (6.36). The result, whose derivation we omit, is given below.

Plate:

$$\frac{T(x,\,t)}{T_c} = \sum_{m=1}^{\infty} \left[\frac{2 \sin \delta_m}{\delta_m + \sin \delta_m \cos \delta_m} \exp\left(-\delta_m^2 \; \text{Fo}\right) \times \cos \frac{\delta_m x}{X} \right]$$

(6.38)

where $\delta \tan \delta = \text{Bi}$ and $\text{Bi} = hX/k$, X denoting the half-thickness of the plate, and $\text{Fo} = at/X^2$.

Cylinder: with radial coordinate denoted by r and external radius by R

$$\frac{T(r,\,t)}{T_c} = \sum_{m=1}^{\infty} \left\{ \frac{2J_1(\mu_m)}{\mu_m \left[J_0^2(\mu_m) + J_1^2(\mu_m) \right]} \exp\left(-\mu_m^2 \text{ Fo}\right) \times J_0\left(\frac{\mu_m r}{R}\right) \right\}$$

(6.39)

Here

$$\mu J_1(\mu) = J_0(\mu) \times \text{Bi}$$

with $\quad\quad\quad \text{Bi} = \dfrac{hR}{k} \quad$ and $\quad \text{Fo} = \dfrac{at}{R^2}$

Sphere of radius R:

$$\frac{T(r,\,t)}{T_c} = \sum_{m=1}^{\infty} \left[2 \frac{\sin \nu_m - \nu_m \cos \nu_m}{\nu_m - \sin \nu_m \cos \nu_m} \exp\left(-\nu_m^2 \text{ Fo}\right) \times \frac{\sin (\nu_m r/R)}{\nu_m r/R} \right]$$

(6.40)

with $\quad\quad\quad\quad\quad \nu \cot \nu = 1 - \text{Bi}$

The forms of the preceding equations conform to the assertions of dimensional analysis, because Eqs. (6.39)–(6.40) agree with the form of Eq. (6.18). This is due to the fact that in the present cases there exists a single temperature T_c or, more precisely, a single temperature difference $T_c - T_\infty$ in the boundary conditions. The eigenvalues δ, μ, and ν depend on the Biot number (Bi) alone. Finally, as expected, the equations are in the form of products stipulated in Eq. (6.10).

Numerical evaluations of Eqs. (6.38)–(6.40) as well as of their associated eigenvalues δ, μ, and ν as functions of the Biot number are available in [6.1] and [6.2]. In practical applications we are most often interested in the temperature at the center, T_m/T_c at $x = 0$ or $r = 0$, and in the wall temperature T_w/T_c at $x = X$ and $r = R$. The equations can be also employed to calculate the quantity of heat Q exchanged in a given time t; it is then useful to normalize this quantity with a properly defined enthalpy difference, Q_c, of the body. The temperature of the surrounding fluid is arbitrarily set equal to zero after its jump-like change. All other temperatures in question, namely T, T_m, T_w, and T_c are measured with respect to this zero level; they may be positive or negative. As $t \to \infty$, the temperature of the body approaches the zero value of the surrounding fluid.

The diagram in Fig. 6.6 describes the center temperature T_m of several bodies for Bi $= \infty$ (boundary condition of the first kind). A sphere cools fastest; the plate cools slowest. For purposes of comparison, we have introduced the curves for several other shapes.

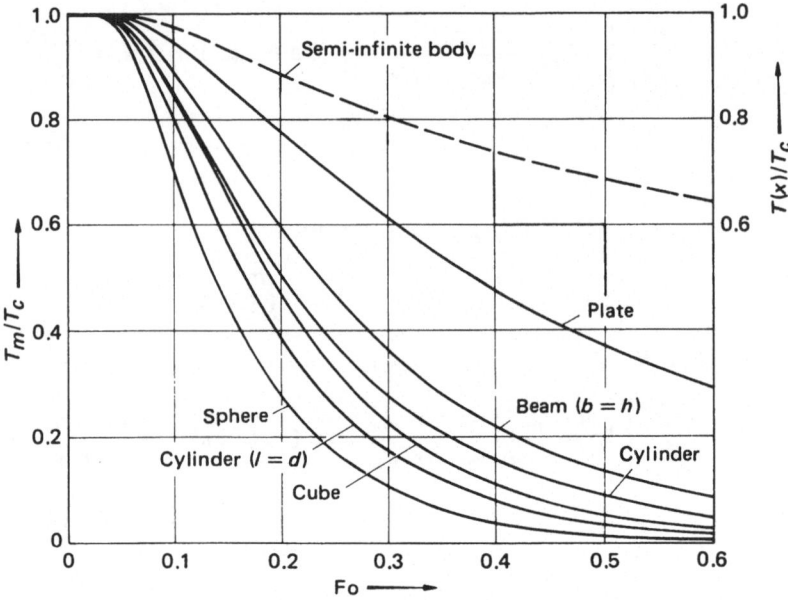

Figure 6.6 Temperature variation at the center of bodies of different geometric shapes and in the interior of a semi-infinite body for Bi → ∞ (constant wall temperature).

Equation (6.38) contains in it the solution for a plate whose one face is adiabatically insulated, because in the plane of symmetry $(x = 0)$ the condition $\partial T/\partial x = 0$ is always satisfied. In this case, X denotes the whole plate thickness.

Example 6.5 A fire wall is to be so designed that it should prevent an increase in temperature on the "cold" side by more than 140 K after a specified time interval upon being suddenly raised in temperature by 700 K on the "hot" side (Fig. 6.7).

The problem is to determine the wall thickness X that would safeguard these conditions during 3, 6, and 12 h, respectively. This is a case of heating, which we can comfortably solve with the aid of the formulas for sudden cooling to $T_\infty = 0$. To achieve this, as shown in Fig. 6.7, it suffices to measure the temperatures T_m and T_c "downward" from the arbitrarily assumed level, $T_\infty = 0$, of the temperature of the fluid. We treat the cold wall as if it were adiabatic, which is a more severe condition than that prevailing in reality, and this allows us to make use of Eq. (6.38) at $x = 0$ and with Bi = ∞. For $T_m/T_c = 560/700 = 0.8$ we get, from Fig. 6.6,

$$\text{Fo} = \frac{at}{X^2} = 0.185$$

where X is the wall thickness. Assuming $a = 0.5 \times 10^{-6}$ m²/s, we calculate

$$X = \left[\frac{at}{\text{Fo}}\right]^{1/2} = 0.17 \text{ m} \qquad \text{for } t = 3 \text{ h}$$

$$X = 0.24 \text{ m} \qquad\qquad \text{for } t = 6 \text{ h}$$

$$X = 0.34 \text{ m} \qquad\qquad \text{for } t = 12 \text{ h}$$

6.7 APPROXIMATE SOLUTIONS

The series of Eqs. (6.38)–(6.40) converge rapidly, especially thanks to the exponential term exp $(-\delta_m^2 \text{Fo})$, because the differences $\delta_{m+1} - \delta_m$ are of order π. It is, therefore, clear that there must exist ranges of values of Bi and Fo for which it is sufficient to retain only the first term of each series. Tolerating an absolute error of about 0.01 in T_m/T_c, T_w/T_c and Q/Q_c, it is possible to use the one-term approximation for any value of Bi and for Fo > Fo*, where Fo* = 0.24 in the case of a flat plate, Fo* = 0.21 in the case of the cylinder, and Fo* = 0.18 for the sphere. The corresponding equations are then:

$$\frac{T_m}{T_c} = C_m \exp(-E \text{ Fo}) \tag{6.41}$$

$$\frac{T_w}{T_c} = C_w \exp(-E \text{ Fo}) \tag{6.42}$$

$$\frac{Q}{Q_c} = 1 - C_q \exp(-E \text{ Fo}) \tag{6.43}$$

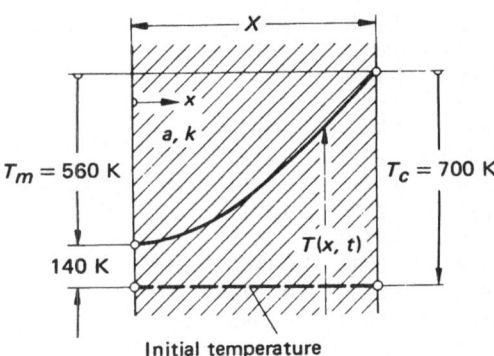

Figure 6.7 Fire wall with adiabatic external face.

The reference enthalpy Q_c is assumed here as

$$Q_c = \rho c_p V T_c$$

where $V = 2XA$ for the plate

$V = \pi R^2 H$ for a cylinder of height H

$V = \frac{4}{3}\pi R^3$ for the sphere

The temperatures at any position x/X or r/R are given by the following approximate formulas:

Plate: $$\frac{T}{T_c} = C_m \exp(-E \, \mathrm{Fo}) \times \cos \frac{\delta_1 x}{X} \qquad (6.44)$$

Cylinder: $$\frac{T}{T_c} = C_m \exp(-E \, \mathrm{Fo}) \times J_0 \frac{\mu_1 r}{R} \qquad (6.45)$$

Sphere: $$\frac{T}{T_c} = C_m \exp(-E \, \mathrm{Fo}) \times \frac{\sin \nu_1 r/R}{\nu_1 r/R} \qquad (6.46)$$

The factors C_m, C_w, C_q, and E depend on the Biot number alone. The same is true about the first eigenvalues δ_1, μ_1, and ν_1; they are all listed in Tables 6.2 to 6.4 for the three simple geometrical shapes under discussion.

The series converge slowly for small values of Fo. In such cases it is possible to recommend special series expansions which the reader can find

Table 6.2 Auxiliary numerical data for Eqs. (6.41)–(6.44) for the flat plate

$\mathrm{Fo} = at/X^2$			$\mathrm{Bi} = hX/k$	$\mathrm{Fo}^* = 0.24$	
$1/\mathrm{Bi}$	C_m	E	C_w	C_q	δ_1
0	1.2732	2.4674	0.0000	0.8106	1.5708
0.1	1.2620	2.0417	0.1785	0.8743	1.4289
0.2	1.2402	1.7262	0.3152	0.9130	1.3138
0.5	1.1784	1.1560	0.5587	0.9635	1.0769
0.8	1.1379	0.8663	0.6796	0.9806	0.9308
1	1.1191	0.7402	0.7299	0.9861	0.8603
2	1.0701	0.4268	0.8498	0.9947	0.6533
5	1.0311	0.1874	0.9360	0.9992	0.4328
8	1.0199	0.1200	0.9594	0.9997	0.3464
10	1.0161	0.0968	0.9673	0.9998	0.3111
20	1.0082	0.0492	0.9835	0.9999	0.2218
50	1.0033	0.0199	0.9934	1.0000	0.1410
80	1.0021	0.0124	0.9958	1.0000	0.1116
100	1.0017	0.0100	0.9967	1.0000	0.0998

Table 6.3 Auxiliary numerical data for Eqs. (6.41)–(6.43) and (6.45) for the cylinder

$Fo = at/R^2$		$Bi = hR/k$		$Fo^* = 0.21$	
1/Bi	C_m	E	C_w	C_q	μ_1
0	1.6020	5.7840	0.0000	0.6916	2.4048
0.1	1.5678	4.7524	0.1905	0.8037	2.1795
0.2	1.5029	3.9601	0.3452	0.8721	1.9898
0.5	1.3386	2.5600	0.6096	0.9536	1.5994
0.8	1.2461	1.8656	0.7293	0.9772	1.3659
1	1.2068	1.5750	0.7764	0.9843	1.2558
2	1.1141	0.8836	0.8812	0.9955	0.9408
5	1.0482	0.3795	0.9510	0.9992	0.6170
8	1.0306	0.2421	0.9691	0.9997	0.4923
10	1.0245	0.1945	0.9753	0.9998	0.4417
20	1.0124	0.0986	0.9876	0.9999	0.3143
50	1.0050	0.0396	0.9950	1.0000	0.1995
80	1.0031	0.0246	0.9969	1.0000	0.1579
100	1.0025	0.0199	0.9975	1.0000	0.1412

Table 6.4 Auxiliary numerical data for Eqs. (6.41)–(6.43) and (6.46) for the sphere

$Fo = at/R^2$		$Bi = hR/k$		$Fo^* = 0.18$	
1/bi	C_m	E	C_w	C_q	v_1
0	2.0000	9.8696	0.0000	0.6079	3.1416
0.1	1.9249	8.0446	0.2040	0.7607	2.8363
0.2	1.7870	6.6071	0.3758	0.8533	2.5704
0.5	1.4793	4.1158	0.6540	0.9534	2.0288
0.8	1.3313	2.9430	0.7679	0.9785	1.7155
1	1.2732	2.4674	0.8106	0.9855	1.5708
2	1.1441	1.3585	0.9021	0.9960	1.1656
5	1.0592	0.5765	0.9603	0.9993	0.7593
8	1.0372	0.3658	0.9751	0.9997	0.6048
10	1.0298	0.2941	0.9801	0.9998	0.5423
20	1.0150	0.1485	0.9900	1.0000	0.3854
50	1.0060	0.0598	0.9960	1.0000	0.2445
80	1.0037	0.0374	0.9975	1.0000	0.1934
100	1.0030	0.0299	0.9980	1.0000	0.1730

listed in [6.2]. Often it suffices to employ the results for the semi-infinite body, namely Eq. (6.19) and the ones following it.

In the more general case, when a semi-infinite body is allowed to cool according to boundary conditions of the third kind, the temperature field is described by the equation

$$\frac{T}{T_c} = \text{erf}\left(\frac{1}{2\,\text{Fo}^{1/2}}\right) + \exp\,(\text{FoBi}^2 + \text{Bi})\,\text{erfc}\left(\frac{1}{2\,\text{Fo}^{1/2}} + \text{Fo}^{1/2}\,\text{Bi}\right)$$

$$(6.47)$$

The dimensionless numbers $\text{Fo} = at/x^2$ and $\text{Bi} = hx/k$ are formed with the distance x from the surface. At $x = 0$ we have $\text{Bi} = 0$ and $1/\text{Fo} = 0$. However, the product

$$\eta = \text{Fo}^{1/2}\,\text{Bi} = \frac{h(at)^{1/2}}{k} = \frac{ht^{1/2}}{b}$$

remains finite, because the geometric distance x has been suppressed in it. For this reason, the temperature at the wall is given by the following simplified form of Eq. (6.47):

$$\frac{T}{T_c} = \exp\,\eta^2 \times \text{erfc}\,\eta$$

The numerical evaluation becomes difficult in cases when the argument

$$z = \frac{1}{2\,\text{Fo}^{1/2}} + \eta$$

in erfc z exceeds the value 2. In such cases it is preferable to make use of the truncated asymptotic expansion

$$\exp z^2 \times \text{erfc}\,z \approx \frac{1}{\pi^{1/2}z}\left(1 - \frac{1}{2z^2} + \frac{3}{4z^4}\right) = \phi(z)$$

so that

$$\frac{T}{T_c} = \text{erf}\left(\frac{1}{2\,\text{Fo}^{1/2}}\right) + \exp\left(-\frac{1}{4\,\text{Fo}}\right) \times \phi\left(\frac{1}{2\,\text{Fo}^{1/2}} + \text{Fo}^{1/2}\,\text{Bi}\right)$$

In order to form a first impression of the behavior of a particular system it is often sufficient to know the functions Q/Q_c or T_m/T_c which occur after a fixed time. A diagram of Fo vs. Bi, provided with lines Q/Q_c treated as a parameter, will be found very useful. Such diagrams are shown in Fig. 6.8 for the plate, in Fig. 6.9 for the cylinder, and in Fig. 6.10 for the sphere.

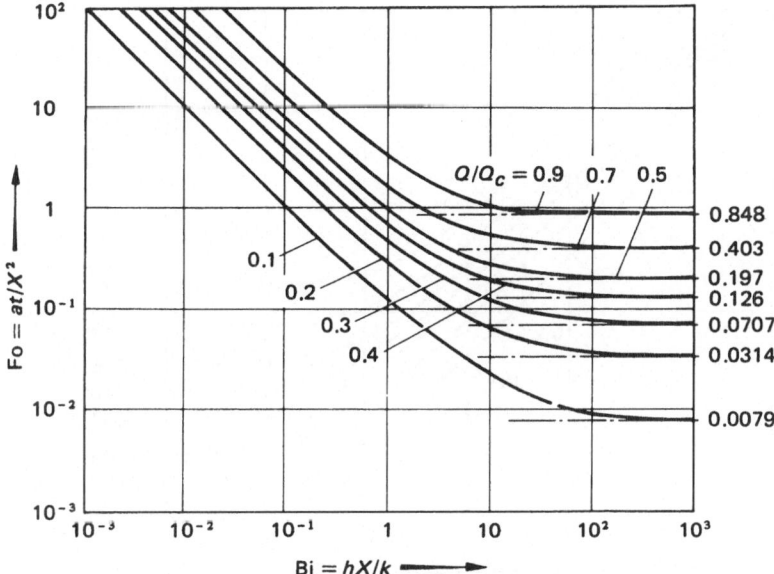

Figure 6.8 The diagram of Fo vs. Bi with Q/Q_c as parameter. The case of the flat plate. The numbers to the right represent asymptotic values for Fo for Bi $\to \infty$.

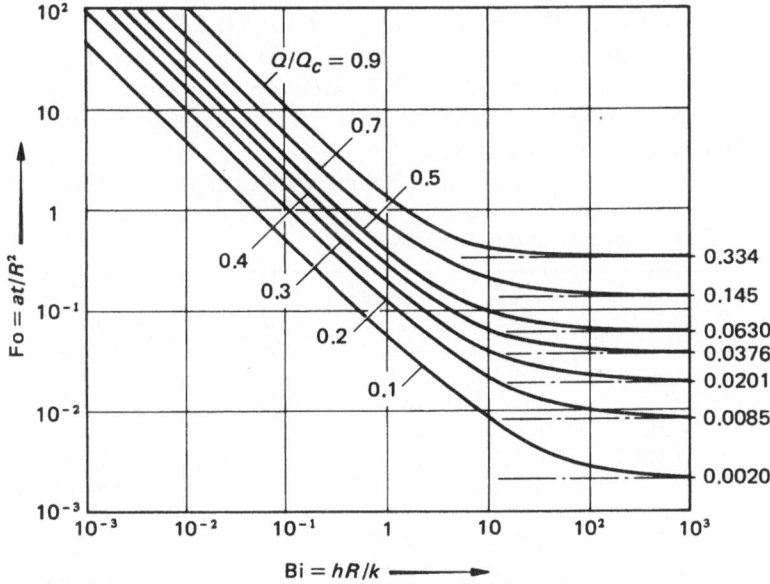

Figure 6.9 The diagram of Fo vs. Bi with Q/Q_c as parameter. The case of the cylinder. See also Fig. 6.8.

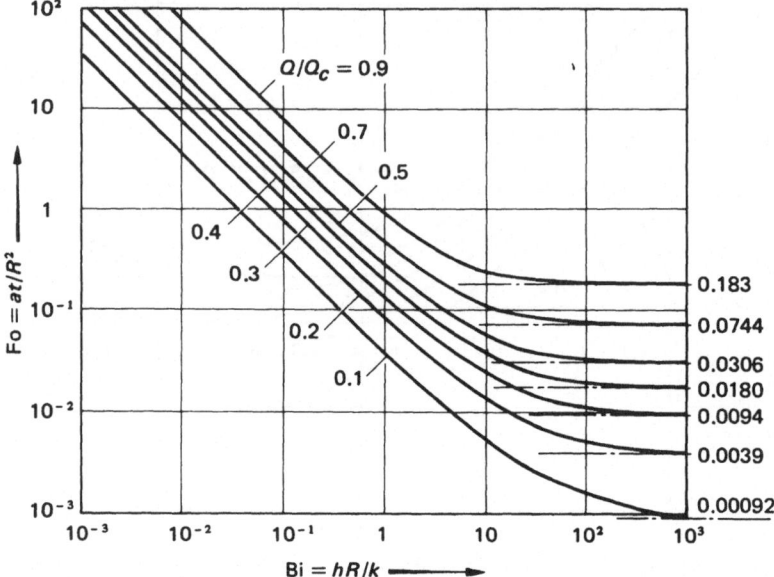

Figure 6.10 The diagram of Fo vs. Bi with Q/Q_c as parameter. The case of the sphere. See also Fig. 6.8.

Example 6.6 We consider three plates of thickness $2X = 0.1$ m made of different materials. The heat transfer coefficient on their external surfaces has the value $h = 10$ W/m² K. We wish to calculate the time intervals after which the condition $Q/Q_c = 0.5$ is attained. The respective data are listed in Table 6.5.

If such a plate is immersed in boiling water, the film coefficient h

Table 6.5 Times that correspond to $Q/Q_c = 0.5$

Magnitude	Insulating plate	Ceramic plate	Metallic plate	Unit	Remarks
X	0.05	0.05	0.05	m	
h	10	10	10	W/m² K	$Q/Q_c = 0.5$
k	0.05	0.5	50	W/m K	occurs for
a	0.278	0.313	12.5	10^{-6} m²/s	different
Bi	10	1	0.01	—	values of Bi
Fo	0.27	0.92	69	—	
t	0.674	2.04	3.83	h	
Fo	0.197	0.197	0.197	—	$Q/Q_c = 0.5$
t_∞	29.5	26.2	0.657	min	for Bi $\rightarrow \infty$

becomes very large. In such a case it is possible to calculate with the asymptotic value $\mathrm{Fo} = 0.197$ for $\mathrm{Bi} = \infty$. The corresponding times t_∞ are also listed in Table 6.5.

6.8 METHOD OF FINITE DIFFERENCES

We apply the method to the solution of the Fourier equation (1.2) for the case of a flat plate in which we approximate the two partial derivatives by ratios of differences. To achieve this, we subdivide the spatial coordinate x and the time coordinate t into equal, finite segments, Δx and Δt, respectively. The space coordinate originates at the surface, as shown in Fig. 6.11. We associate with each midpoint a discrete value of temperature that corresponds to time $t = m\,\Delta t$, where m is an integer. The midpoint of the nth spatial interval is located at $x = (n - \frac{1}{2})\,\Delta x$, where n, too, is an integer. The temperature that occurs at that point at $t = m\,\Delta t$ is denoted by the symbol T_{nm}.

The partial derivative of second order with respect to x, which appears

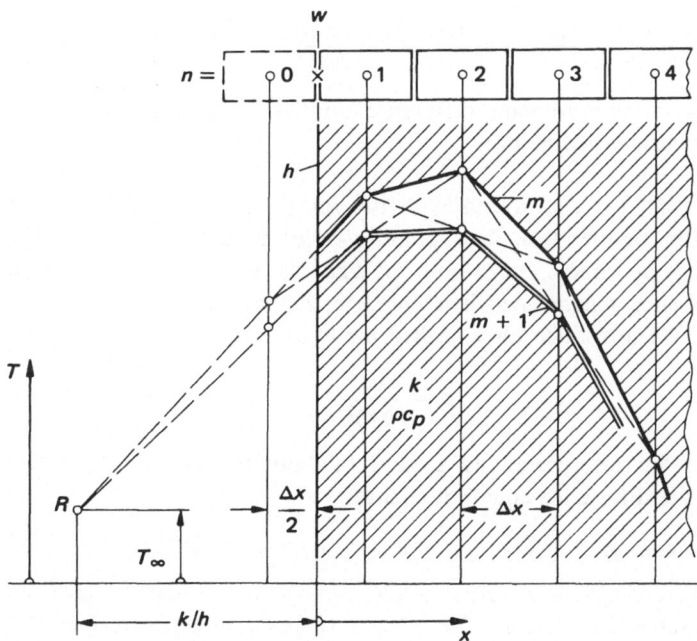

Figure 6.11 Graphical construction for the application of the method of finite differences due to Binder and Schmidt.

in Eq. (1.2) at time t and location x, is approximated by a "central" difference ratio of second order. This can be expressed as the difference between two central difference quotients of first order at locations $x + \frac{1}{2}\,\Delta x = n\,\Delta x$ and $x - \frac{1}{2}\,\Delta x = (n-1)\,\Delta x$, and leads to the approximation

$$\left(\frac{\partial^2 T}{\partial x^2}\right)_{xt} \approx \frac{(T_{(n+1)m} - T_{nm})/\Delta x - (T_{nm} - T_{(n-1)m})/\Delta x}{\Delta x}$$

$$= \frac{T_{(n+1)m} - 2T_{nm} + T_{(n-1)m}}{(\Delta x)^2}$$

The partial derivative of first order with respect to time appearing in Eq. (1.2) will be approximated by the "forward" difference quotient

$$\left(\frac{\partial T}{\partial t}\right)_{xt} \approx \frac{T_{n(m+1)} - T_{nm}}{\Delta t}$$

It is immediately evident that it would be better to interpret this form as a "central" approximation for time $t + \frac{1}{2}\,\Delta t$; nevertheless, we employ it as a sufficiently good approximation for time t. Substituting the above approximations into Eq. (1.2), we obtain the following difference equation:

$$\frac{T_{n(m+1)} - T_{nm}}{\Delta t} = a\,\frac{T_{(n+1)m} - 2T_{nm} + T_{(n-1)m}}{(\Delta x)^2} \qquad (6.48)$$

The selection of a forward difference quotient with respect to time entails two essential consequences:

(a) First, Eq. (6.48) can be solved explicitly for $T_{n(m+1)}$. This allows us to evaluate the temperature field point-by-point in the x direction and step-by-step in time. The result is what is known as an explicit finite-difference scheme which is characterized by utmost simplicity of execution. We introduce the so-called modulus

$$p = \frac{a\,\Delta t}{(\Delta x)^2} \qquad (6.49)$$

which can be interpreted as a kind of Fourier number. With this notation, the difference equation changes to

$$T_{n(m+1)} = p(T_{(n+1)m} + T_{(n-1)m}) + (1 - 2p)T_{nm} \qquad (6.50)$$

(b) The second consequence—rather unpleasant—is that p may be given at most the value 0.5, because for $p > 0.5$ there occur negative numbers in Eq. (6.50), and this completely destabilizes the calculation. This first condition of stability puts an upper limit

$$\Delta t \leqslant \frac{1}{2} \frac{(\Delta x)^2}{a} \tag{6.51}$$

on the choice of the time interval that can be made for a prescribed value of Δx. In cases when a fine mesh Δx is selected, the limitation on Δt may force us to employ a very large number of steps in time.

If the modulus p is given its limiting value 0.5, Eq. (6.50) loses its last term and the temperature at time $m + 1$ at location n becomes exactly equal to the arithmetic mean between the temperatures at the neighboring locations $n - 1$ and $n + 1$ at time m. This algebraic operation can be performed very simply by "cutting corners," as illustrated in Fig. 6.11. This method was first invented by Binder in 1910 who also derived Eq. (6.50); it was independently reinvented by Schmidt in 1924.

The boundary condition of the third kind defined in Eq. (1.8) is the one that occurs most often in problems of heat transfer of practical importance. For this reason, we shall now transform it into a finite-difference equation. Employing the notation of Fig. 6.11, we may write

$$k\left(\frac{T_{1m} - T_{wm}}{\frac{1}{2}(\Delta x)}\right) = h(T_{wm} - T_{\infty m}) \tag{6.52}$$

The simplest way to incorporate this condition into the calculational scheme is to introduce an auxiliary layer of material. We have assigned the location index 0 to this layer in Fig. 6.11 and we have indicated the layer itself by a broken line. It is stipulated here that the process between points 0 and 1 is one of pure conduction, so that the relation

$$\frac{k}{\Delta x/2}(T_{wm} - T_{0m}) = \frac{k}{\Delta x/2}(T_{1m} - T_{wm}) \tag{6.53}$$

must be satisfied. We introduce the special Biot number

$$\mathrm{Bi}^* = \frac{h\,\Delta x}{2k} \tag{6.54}$$

and eliminate the intermediate temperature T_{wm} in favor of the temperature in the fictitious layer. This yields

$$T_{0m} = \frac{1 - \mathrm{Bi}^*}{1 + \mathrm{Bi}^*} T_{1m} + \frac{2\,\mathrm{Bi}^*}{1 + \mathrm{Bi}^*} T_{\infty m} \tag{6.55}$$

The adoption of this fictitious temperature has transformed the influence of the surroundings on the solid body into a pure conduction process from location 0 to location 1, and this makes it possible to use Eq. (6.50) for $n = 1$ thus providing the value at point 0 as base point without introducing a contradiction. This extension can be introduced geometrically with great ease: After determining temperatures $T_{1(m+1)}$, $T_{2(m+1)}$, $\ldots$, etc., by "cutting corners," we replace subscript $m + 1$ by m. In other words, the sequence of values that has just been determined is regarded as a starting series for the next step in time. Before we start the new calculation of the temperature field, it is necessary to determine the temperature in the auxiliary layer, because application of the difference equation (6.50) to location $n = 1$ is based on this value as its base point. To do this we draw a straight line between the "directional point" R placed at a distance k/h from the face (see Sec. 1.3) and the temperature T_{1m} of the inner layer closest to the external face. The point of intersection with the auxiliary ordinate at $x = -\frac{1}{2} \Delta x$, i.e., at $n = 0$, yields the required temperature T_{0m}. This geometrical construction exactly satisfies boundary condition (6.55). Equation (6.53) yields the value $T_{wm} = \frac{1}{2}(T_{0m} + T_{1m})$ for the temperature at the external wall. If a similar boundary condition is prescribed on the other external face (righthand face in Fig. 6.11), it becomes necessary to add an auxiliary layer there too; this is located at $x = X + \frac{1}{2} \Delta x$ and acquires subscript $n + 1$. The remaining procedure is analogous to the one just explained.

If the Biot number from Eq. (6.54) were to assume values exceeding unity (R locates itself then between $x = 0$ and $x = -\frac{1}{2} \Delta x$), the calculation could show evidence of instability at point $n = 1$, because this violates a second stability criterion. We substitute boundary condition (6.55) into the difference equation (6.50) for $n = 1$ and require that no coefficient shall become negative. In this manner, after some transformations, we are led to the criterion that

$$p \leqslant \frac{1 + \text{Bi}^*}{1 + 3\,\text{Bi}^*} \tag{6.56}$$

It is easy to verify by substitution into Eq. (6.56) that for $\text{Bi}^* > 1$ we obtain the more stringent stability condition $p < \frac{1}{2}$. The limiting case with $\text{Bi}^* \to \infty$ puts on p the condition $p \leqslant \frac{1}{3}$. It follows that the construction is stable for all values of Bi^* if the condition $p \leqslant \frac{1}{3}$ is maintained. However, the possibility of using the very simple procedure of "cutting corners" is lost. Nowadays, with powerful pocket calculators, this is no longer a significant drawback. Moreover, a numerical calculation in tabular form is considerably more accurate and the ability to choose a value for the modulus p which is much lower than the limit in Eq. (6.56) lends greater flexibility (conduction through layered structures with layers of different thermophysical properties).

The preceding Binder-Schmidt method can easily be applied to the solution of nonlinear problems in heat conduction (e.g., when h is a function of wall temperature) or to problems in which there occurs a variable external temperature. In either case it is found that the directional point R is displaced along a suitably determined curve. The method can also be extended to apply to cylindrical or spherical geometries. In all cases the solution of problems with arbitrarily distributed heat sources can be performed without undue effort. A good introduction into this set of problems can be found in [6.3].

Example 6.7 A flat plate of thickness $X = 0.1$ m and initial temperature $T_c = 100°C$ is insulated adiabatically on one face and exposed to an air stream of temperature $T_\infty = 0°C$ on the other. The properties of the plate are: $k = 25$ W/m K and $a = 7.1 \times 10^{-6}$ m^2/s; the coefficient of heat transfer is $h = 250$ W/m^2 K. We divide the plate into $n = 3$ segments and choose the module value $p = 0.5$. We are to determine the discrete series describing the temperature distribution after $m = 6$ time steps. Since $p = 0.5$ allows us to use the method of "cutting corners," it is possible to employ the graphical construction as a check on the calculation. We shall, however, dispense with this possibility here.

Choosing $\Delta x = \frac{1}{3} X = 0.0333$ m, we calculate $\Delta t = 78.2$ s from Eq. (6.49). The total cooling time of interest is thus $t = 6\,\Delta t = 470$ s. The special Biot number from Eq. (6.54) becomes $Bi^* = 1/6$. In the presence of two boundary conditions, we must add two fictitious layers. The auxiliary layer at the adiabatic face will be given the number 0, and the other the ordinal number $n + 1 = 4$. At the adiabatic face we have $h = 0$, that is $Bi^* = 0$, which constitutes a special case of a boundary condition of the third kind. This means that Eq. (6.55) can still be applied for the calculation of the auxiliary temperature T_{0m}. Equation (6.55) yields T_{4m} after T_{3m} has been substituted in it for T_{1m}. This leads to

$$T_{0m} = T_{1m} \qquad \text{for } Bi^* = 0$$

$$T_{4m} = \tfrac{5}{7} T_{3m} \qquad \text{for } Bi^* = \tfrac{1}{6}$$

The temperatures $T_{1(m+1)}$ to $T_{3(m+1)}$ are calculated from Eq. (6.50). In view of the assumed value $p = 0.5$, the latter simplifies to

$$T_{n(m+1)} = \tfrac{1}{2}(T_{(n+1)m} + T_{(n-1)m})$$

We first calculate the inner temperatures; the temperatures at the faces are calculated only afterward. The course of the calculation can be understood in detail by reference to Table 6.6. The reduced heat Q/Q_c, lost after m steps, is calculated with the aid of the relation

Table 6.6 Schema for calculation by use of the method of finite differences with $p = 0.5$
Last line contains the results of a check-calculation

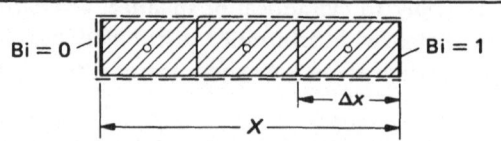

m	T_{0m}/T_C	T_{1m}/T_C	T_{2m}/T_C	T_{3m}/T_C	T_{4m}/T_C	$(Q/Q_C)_m$
0	1.000	1.000	1.000	1.000	0.714	0
1	1.000	1.000	1.000	0.857	0.612	0.048
2	1.000	1.000	0.929	0.806	0.576	0.088
3	0.965	0.965	0.903	0.752	0.537	0.127
4	0.934	0.934	0.859	0.720	0.514	0.162
5	0.897	0.897	0.827	0.687	0.490	0.196
6	0.862	0.862	0.792	0.659	0.470	0.229
6	–	0.865	0.795	0.659	–	0.230

$$\left(\frac{Q}{Q_c}\right)_m = \frac{1}{3\,T_c} \sum_{n=1}^{3} (T_c - T_{nm})$$

The corresponding numerical values are given in the last column of the table.

The accuracy of the calculation can be checked with reference to the approximate solutions explained in Sec. 6.7, since

$$\mathrm{Fo} = \frac{at}{X^2} = \frac{mp}{n^2} = \frac{1}{3} > \mathrm{Fo}^* = 0.24$$

For $\mathrm{Bi} = hX/k = 1$, Table 6.2 gives

$$\delta = 0.860 \qquad C_m = 1.119 \qquad E = 0.740 \qquad C_q = 0.986$$

In our problem

$$\frac{T_{n6}}{T_c} = C_m \exp\left(-E\,\mathrm{Fo}\right) \cos\left(\delta_1 \frac{n - 0.5}{3}\right)$$

The numerical results, together with Q/Q_c, calculated with the aid of Eq. (6.43), are listed in the last line of Table 6.6. Bearing in mind the crudeness of the division into $n = 3$ layers, we conclude that the agreement with the almost exact approximate equation is surprisingly good. It is plausible to suppose that the even cruder scheme with $n = 2$ might give useful results.

6.9 EXPERIMENTAL ANALOGS

We propose to present here three nonsteady transport processes chosen from among quite diverse branches of physics. The mathematical description of each process leads to the same parabolic differential equation (1.2) as the one which Fourier derived for the transport of heat. In cases like these, we speak of analogous processes. We have here in mind the following:

1. The conduction of an electric current in a cable produced by a variable applied voltage
2. The laminar seepage through an earthen dam in the presence of a variable head, for example, due to the ebb and flow of the tide
3. The diffusion of extraneous atoms in a semiconductor crystal

In the case of multidimensional transport processes with complex boundaries or nonlinear boundary conditions, it is very rare to be in a position to find analytic solutions to the problem. For this reason, many research workers have made attempts to solve such problems with the aid of so-called analog models. The results obtained with the aid of such substitute systems can then be applied to, or translated to, each of the above fields on the basis of the existing analogy.

The simplest to construct are electrical or hydraulic models. In either case, we retain the continuous variation with time, but the quantities that describe the geometry of the system are modeled with the aid of discrete resistors or capacitors. In other words, the continuum is divided up into individual, in most cases equal, building blocks which can be readily bought or made.

6.9.1 The Electric Analog (Beuken, 1936)

In order to explain the principle of construction of an electric model, we show in Fig. 6.12 an electric network that allows us to simulate the flow of heat across a flat plate. The plate of surface area A with thickness X and material properties k (thermal conductivity), ρ (density), c_p (specific heat) is subdivided into $n = 4$ slabs of equal thickness Δx. A film coefficient h is prescribed on one face (boundary condition of the third kind), whereas the other face is assumed to be adiabatic.

The building blocks of the electric model consist of ohmic resistors R (SI unit: ohm, Ω) and condensers of capacity C (SI unit: farad, F). The potential is given by the electric voltage U (SI unit: volt, V). Introducing the analog proportionality factors $\mu_1, \ldots, \mu_4$, we can establish the following relations between the electric and the thermal quantities:

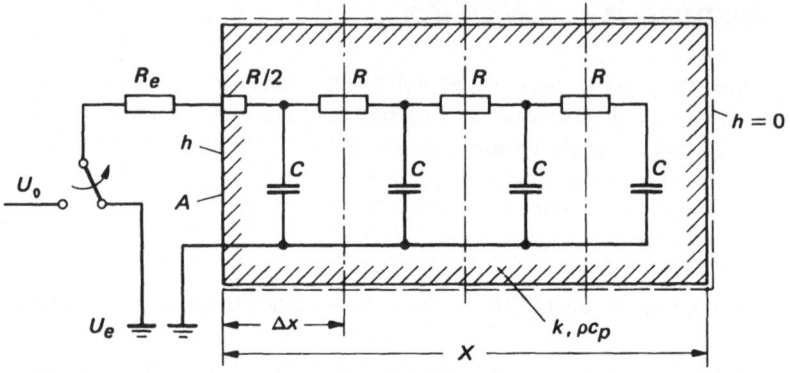

Figure 6.12 Schematic diagram of an electric analog model.

$$R = \mu_1 \frac{\Delta x}{kA} \qquad \text{(internal resistance)} \qquad (6.57)$$

$$R_e = \frac{\mu_1}{hA} = \frac{Rk}{h\,\Delta x} \qquad \text{(external resistance)} \qquad (6.58)$$

$$C = \mu_2 \rho c_p A\,\Delta x \qquad \text{(capacitance)} \qquad (6.59)$$

$$t_e = \mu_3 t \qquad \text{(time)} \qquad (6.60)$$

$$U = \mu_4 T \qquad \text{(voltage, i.e., potential)} \qquad (6.61)$$

Basing oneself on the considerations of Sec. 3.3, we can now easily demonstrate that the discharging of a thermal storage unit of capacity $\rho c_p A\,\Delta x$ through a thermal resistance of magnitude $\Delta x/kA$ is governed by the equation

$$\frac{T - T_e}{T_0 - T_e} = \exp\left(-\frac{kt}{\rho c_p (\Delta x)^2}\right) \qquad (6.62)$$

for the temperature drop. Here T_e denotes the external temperature and T_0 is the initial heat-reservoir temperature.

The potential decrease in a condenser C which discharges through a resistor R is similarly described by the equation

$$\frac{U - U_e}{U_0 - U_e} = \exp\left(-\frac{t_e}{RC}\right) \qquad (6.63)$$

Introducing the relations in Eqs. (6.57) and (6.59)–(6.61) into Eq. (6.63), we are led to

$$\frac{T - T_e}{T_0 - T_e} = \exp\left(-\frac{\mu_3}{\mu_1 \mu_2} \frac{kt}{\rho c_p (\Delta x)^2}\right) \qquad (6.64)$$

The required condition for the transfer of results from the electric model to the thermal original can be based on Eqs. (6.62) and (6.64) and expressed as follows:

1. The transfer coefficient μ_4 can be given an arbitrary value because we are interested only in relative variations in the potential.
2. When the resistors and condenser have fixed values—that is, in the face of prescribed transfer coefficients μ_1 and μ_2—we find that the transfer coefficient for time is

$$\mu_3 = \mu_1\mu_2 \tag{6.65}$$

It then follows that the times of the two systems are in the ratio

$$\frac{t_e}{t} = RC\frac{k}{\rho c_p(\Delta x)^2} \tag{6.66}$$

3. According to Eq. (6.58) and Fig. 6.12, the total external resistance must have the value

$$R_{e,\text{tot}} = R\left(\frac{k}{h\,\Delta x} + \frac{1}{2}\right)$$

6.9.2 The Hydraulic Model
(Moore, 1935; Lukyanov, 1936)

Figure 6.13 depicts a hydraulic analog model which is also capable of simulating the process of heat penetration into a flat plate. Transparent tubes of diameter D form the storage containers and capillary tubes of diameter d and length l act as resistances. Both pipe systems are filled with a fluid (colored water in most cases) of density ρ_h and viscosity μ_h. Scales

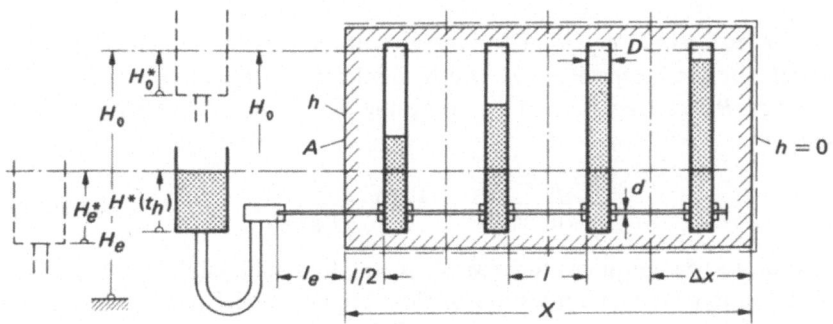

Figure 6.13 Schematic diagram of a hydraulic analog model.

placed behind the standing pipes readily indicate the height H and hence the local values of the potential; these can be easily read off or photographed.

The preceding section has demonstrated that three coefficients of proportionality are sufficient to establish relations between the hydraulic and the thermal quantities. Further, Eq. (6.58) demonstrates that the re-creation of an external resistance presents no new problems after the internal resistances have been properly constructed.

According to the law of Hagen and Poiseuille, the pressure drop Δp suffered by a stream of intensity $\dot{m}$ flowing through a capillary in the laminar regime is given by

$$\Delta p = \frac{128}{\pi d^4} \frac{\mu_h}{\rho_h} l\dot{m} \tag{6.67}$$

If this pressure drop is to be modeled by the height H of a column of liquid, it is necessary to satisfy the condition that

$$H = \frac{\Delta p}{g \rho_h} = \frac{128 l}{\pi d^4 g} \frac{\mu_h}{\rho_h^2} \dot{m} \tag{6.68}$$

where g denotes the acceleration due to gravity. The preceding equation allows us to define a unit internal hydraulic resistance R_h:

$$R_h = \frac{128 l}{\pi d^4 g} \frac{\mu_h}{\rho_h^2} = \mu_1 \frac{\Delta x}{kA} \tag{6.69}$$

We, further, define the capacity C_h of a vertical tube as

$$C_h = \rho_h \frac{\pi D^2}{4} = \mu_2 \rho c_p A \, \Delta x \tag{6.70}$$

and set the time ratio, in analogy with Eq. (6.60), as

$$t_h = \mu_3 t$$

The differential equation for the variation in the level of the fluid in a vertical tube of capacity C_h which discharges through a capillary of resistance R_h can be derived by using the law of mass conservation. The solution is

$$\frac{H - H_e}{H_0 - H_e} = \exp\left(-\frac{t_h}{R_h C_h}\right) \tag{6.71}$$

Equation (6.65) continues to be valid for the hydraulic model ($\mu_3 = \mu_1 \mu_2$), and the ratio of the two measures of time becomes equal to

$$\frac{t_h}{t} = \frac{128 l}{\pi d^4 g} \frac{\mu_h}{\rho_h} \frac{\pi D^2}{4} \frac{k}{\rho c_p (\Delta x)^2} \tag{6.72}$$

The mass of fluid that entered (or left) up to an instant t is proportional to the quantity of heat transferred through the appropriate face of the thermal model. In this connection, it is useful to employ graduated movable bottles (shown at the left in Fig. 6.13). If all vertical tubes are filled to the same maximum height H_0, then the corresponding level in the movable cylinder is denoted by H_0^*. If, on the other hand, the former are filled to the minimum level H_e (this is the assumed value of the external potential that exists in the equalization process), the corresponding level of the movable cylinder will be denoted by H_e^*. Thus, the relative quantity of heat transferred becomes

$$\frac{Q(t)}{Q_c} = \frac{H_e^* - H^*(t_h)}{H_e^* - H_0^*}$$

The analog models described in this section constitute special designs of analog computers. They may still be used in practice, in spite of the ready availability of electronic simulators in present times, owing to their low construction cost and attractive helpfulness in visualizing the process.

6.10 THE LAPLACE TRANSFORMATION

The Laplace transformation subordinates a new function $F(s)$ to a given function $f(t)$. The prescription is given by the integral

$$F(s) = \int_0^\infty \exp(-st) f(t)\, dt \tag{6.73}$$

Here t denotes the independent variable and s is a new variable, the so-called Laplace variable. The exponential function $\exp(-st)$, which appears in the integrand, is called its kernel. The kernel imposes a strongly damped behavior on the integrand, and for this reason the improper integral converges when calculated for almost all functions $f(t)$ that may occur in practical applications. The function $f(t)$ and $F(s)$ are known, respectively, as the original, or primary function, and as its (Laplace) transform. The Laplace transform maps the function of real, physical coordinates on its subordinate function whose independent variable can assume complex values.

The two functions are distinguished from each other, as was done above, by using a lowercase f for the primary function and an uppercase F for its Laplace transform. An alternative notation consists in using the same symbol, but providing the symbol for the transform with a bar. (This, however, may lead the reader to confuse it with the complex conjugate value for the former!) The relationship between the two functions, defined by the integral transform of Eq. (6.73), is often represented symbolically by the use of the operator notation $\mathcal{L}$, viz.,

$$F(s) = \mathcal{L}\{f(t)\} \qquad \text{or} \qquad \bar{g}(s) = \mathcal{L}\{g(t)\}$$

The inverse transform, that is the operation which retrieves the original function from a prescribed Laplace transform, is expressed by an integral in the complex domain. We refrain from citing it here, as its knowledge is not essential for our purposes. The inverse operation is denoted symbolically by

$$f(t) = \mathcal{L}^{-1}\{F(s)\} \qquad \text{or} \qquad g(t) = \mathcal{L}^{-1}\{\bar{g}(s)\}$$

The relationship between f and F is a kind of correspondence for which the symbol ○—● may be used:

$$f(t) \,○\!\!-\!\!● \, F(s) \qquad \text{or} \qquad \bar{g}(s) \,●\!\!-\!\!○ \, g(t)$$

The application of Laplace transforms is eminently well suited to the solution of ordinary or partial linear differential equations, particularly where systems of such equations occur, on condition that the transformed equation must satisfy an initial-value boundary condition. Many time-dependent problems fall under this heading, and this explains the use of t as the argument in $f(t)$.

The method of Laplace transforms (see [6.4] by Doetsch) can be regarded as a kind of language for which tables of corresponding functions become analogous to a dictionary. These tables list a large number of correspondences of the type $F(s) \,●\!\!-\!\!○ \, f(t)$. The operations that are to be performed, not unlike the rules for differentiation and integration, play the part of the rules of grammar for this language. The correspondences that we shall need later are listed in Table 6.7.

In many cases we are not at all interested in obtaining an explicit expression for the solution of a differential equation. Similarly—and this occurs even more often—the solution contains higher transcendental functions that are very difficult to evaluate. In such cases, use of Laplace transforms provides us with a very simple method of obtaining expansions of the original function around $t = 0$ or so-called asymptotic expansions for $t \to \infty$ directly from its Laplace transform! To achieve this, it suffices to expand $F(s)$ into a series—most often by the binomial theorem—around $s = \infty$ and $s = s_0$, respectively. These series are subsequently transformed back term by term. The value s_0 is then the singularity [in most cases a pole] whose real part is farthest to the right on the real axis.

Example 6.8 At instant $t = 0$, we introduce hot machine parts made of steel into a well-stirred oil bath. We are required to determine the

Table 6.7 Operations and correspondences for the Laplace transformation

Operations	
$f'(t) \circ\!\!-\!\!\bullet sF(s) - f(+0)$ $f''(t) \circ\!\!-\!\!\bullet s^2 F(s) - f(+0)\cdot s - f'(+0)$	Differentiation of original function
$f(at) \circ\!\!-\!\!\bullet F(s/a)/a; a > 0$ $F(as) \bullet\!\!-\!\!\circ f(t/a)/a; a > 0$	Principle of similarity
$\exp(-at)f(t) \circ\!\!-\!\!\bullet F(s + a)$	Damping theorem
$\displaystyle\int_0^t f(\tau)\, d\tau \circ\!\!-\!\!\bullet F(s)/s$	Integration of original function
$\displaystyle\int_0^t f_1(\tau)f_2(t-\tau)\, d\tau \circ\!\!-\!\!\bullet F_1(s)\cdot F_2(s)$	Convolution (or *faltung*) theorem

Correspondences	
a	a/s
$\exp(-at)$	$1/(s+a)$
$[1 - \exp(-at)]/a$	$1/[s(s+a)]$
at	a/s^2
$\cos at$	$s/(s^2 + a^2)$
$\sin at$	$a/(s^2 + a^2)$
$\cosh at$	$s/(s^2 - a^2)$
$\sinh at$	$a/(s^2 - a^2)$
$a/t^{1/2}$	$a\pi^{1/2}/s^{1/2}$
$\mathrm{erf}\,(at)^{1/2}$	$a^{1/2}/[s(s+a)^{1/2}]$
$\exp(a^2 t) \times \mathrm{erfc}\,(at^{1/2})$	$[s^{1/2}(a + s^{1/2})]^{-1}$
$\mathrm{erfc}\,(a/(2t^{1/2}))$	$\exp(-as^{1/2})/s$

temperature variation T_B of the bath as well as of the parts T_S as functions of time. We assume that none of the elements shows local temperature differences, except for thin boundary layers in the oil formed around the steel parts.

We are given the following data:

Initial temperatures: T_{B0} and T_{S0}
Heat capacities: $(mc)_B$ and $(mc)_S$
Total external surface of steel parts: A
Coefficient of heat transfer: h

Heat balances for the oil and the metal parts result in two ordinary, coupled differential equations

$$-(mc)_S \frac{dT_S}{dt} = hA(T_S - T_B)$$

$$(mc)_B \frac{dT_B}{dt} = hA(T_S - T_B)$$

We introduce the following dimensionless variables

$$\Theta = \frac{T - T_{B0}}{T_{S0} - T_{B0}} \qquad \tau = \frac{hAt}{(mc)_S} \qquad \epsilon = \frac{(mc)_S}{(mc)_B} \qquad (6.74)$$

and transform our equations to read

$$\frac{d\Theta_S}{d\tau} + \Theta_S - \Theta_B = 0$$

$$\frac{d\Theta_B}{d\tau} - \epsilon(\Theta_S - \Theta_B) = 0 \qquad (6.75)$$

with the initial conditions

$$\Theta_S(0) = 1 \qquad \Theta_B(0) = 0 \qquad \text{for } \tau = 0$$

Employing the rule of differentiation from Table 6.7, which states that the transform of the derivative of the original function is equal to the product of the transform multiplied by the transform variable minus the initial value of the original function in the real domain (!), we obtain a simplified version of our original equations:

$$s\bar{\Theta}_S - 1 + \bar{\Theta}_S - \bar{\Theta}_B = 0$$

$$s\bar{\Theta}_B - \epsilon\bar{\Theta}_S + \epsilon\bar{\Theta}_B = 0$$

These are now two *algebraic* equations for the transforms $\bar{\Theta}_S$ and $\bar{\Theta}_B$, which are evaluated explicitly as

$$\bar{\Theta}_B = \frac{\epsilon}{s(s + \epsilon + 1)} \qquad (6.76)$$

$$\bar{\Theta}_S = \frac{1}{s + \epsilon + 1} + \frac{\epsilon}{s(s + \epsilon + 1)} \qquad (6.77)$$

The inverse transforms are now found with the aid of the correspondences listed in Table 6.7. They lead us directly to a solution of our problem. Thus

$$\Theta_B = \frac{\epsilon}{1 + \epsilon} \{1 - \exp[-(\epsilon + 1)\tau]\} \qquad (6.78)$$

$$\Theta_S = \exp[-(\epsilon + 1)\tau] + \Theta_B = \frac{\epsilon + \exp[-(\epsilon + 1)\tau]}{\epsilon + 1} \qquad (6.79)$$

If it were desired to obtain approximate solutions for very short or very long times, it would suffice to expand the solutions of Eqs. (6.76) and (6.77) about $s = \infty$ and $s = 0$, respectively, which would yield the following series:

$$\bar{\Theta}_B = \frac{\epsilon}{s^2} - \frac{\epsilon(\epsilon + 1)}{s^2} - \cdots \qquad \text{for } s \to \infty$$

$$\bar{\Theta}_S = \frac{1}{s} - \frac{1}{s^2} - \cdots \qquad \text{for } s \to \infty$$

$$\bar{\Theta}_B = \frac{\epsilon}{s(\epsilon + 1)} + \text{terms with exponents} \geq 0 \qquad \text{for } s \to 0$$

$$\bar{\Theta}_S = \frac{\epsilon}{s(\epsilon + 1)} + \text{terms with exponents} \geq 0 \qquad \text{for } s \to 0$$

In writing down the inverse transforms, we limit the expressions to first-order terms in τ and note that only terms with negative exponents in s in the transform contribute terms of order τ in the original. This leads us to the approximations

$$\Theta_B \approx \epsilon\tau \qquad \text{(tangent at } \tau = 0)$$

$$\Theta_S \approx 1 - \tau \qquad \text{(tangent at } \tau = 0)$$

$$\Theta_B = \Theta_S \approx \frac{\epsilon}{1 + \epsilon} \qquad \text{(horizontal asymptote for } \tau \to \infty)$$

Figure 6.14 contains graphs of the functions (6.78) and (6.79) together with their approximations for $\epsilon = 0.25$. When steel parts are quenched in oil, the parameter ϵ is at most $\epsilon = 0.05$. We have chosen a larger value, because for $\epsilon \leq 0.05$ the diagram of Fig. 6.14 would become less clear.

We introduce further examples of the application of the method of Laplace transforms in the next section.

The standard reference [6.4] and the short introduction in [6.5] can be recommended for further study because they emphasize applications.

6.11 PERIODIC TEMPERATURE VARIATION

We consider a homogeneous, isotropic semi-infinite body of thermal conductivity k and thermal diffusivity a. The temperature distribution in such a body is again described by the Fourier equation (1.2), namely

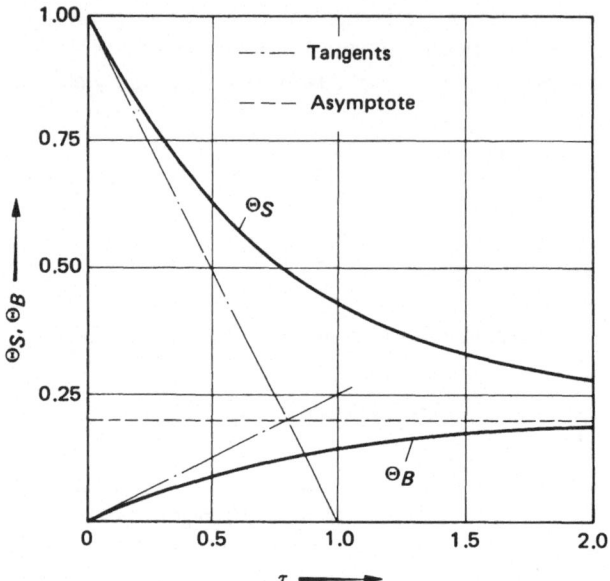

Figure 6.14 Variation of temperature in oil bath and in steel parts represented with the aid of dimensionless variables; $\epsilon = 0.25$.

$$\frac{\partial T}{\partial t} = a \frac{\partial^2 T}{\partial x^2} \tag{6.80}$$

Figure 6.15 shows the positive direction chosen for the coordinate x. We assume that the temperature of the fluid in contact with the free external surface varies periodically according to the relation

$$T_e = T_m + \Delta T \cos \frac{2\pi t}{t_0} \tag{6.81}$$

where T_m is a constant mean temperature, ΔT is the amplitude of the temperature fluctuation, and t_0 is the period. We prescribe a boundary condition of the third kind, so that at the surface the temperature field must satisfy the condition

$$h[T_e - T(0, t)] = -k \left(\frac{\partial T}{\partial x} \right)_{x=0} \qquad \text{for } x = 0 \tag{6.82}$$

where h is the film coefficient. We are to find the temperature distribution $T(x, t)$ in the semi-infinite body as well as the temporal variation of the heat flux, $\dot{q}(0, t)$, at the surface which exists after a stationary state has set in. The latter will establish itself after the lapse of a long time, that is when the transient will have died out. For this reason we may assign an arbitrary value for $T(x, 0)$, and put

$$T(x, 0) = 0 \qquad (6.83)$$

The assumption that the body is semi-infinite leads us to the second boundary condition

$$T(\infty, t) \text{ must be finite for } x \to \infty \qquad (6.84)$$

We calculate the Laplace transforms[†] of Eqs. (6.80)–(6.84), and use the data of Table 6.7 to obtain

$$s\bar{T} = a \frac{d^2 \bar{T}}{dx^2}$$

$$x = 0: \quad h\left[\frac{T_m}{s} + \frac{s \, \Delta T}{s^2 + 4\pi^2/t_0^2} - \bar{T}(0, s)\right] = -k\left(\frac{d\bar{T}}{dx}\right)_{x=0}$$

$$x \to \infty: \quad \bar{T}(\infty, s) \text{ finite}$$

[†]The problem under consideration could also have been solved by the use of Duhamel's theorem. This would have forced us to evaluate very complicated, definite integrals. For this reason, we have preferred to use the method of Laplace transforms and so to take advantage of its highly schematized procedures. The transforms are taken with respect to t and not x.

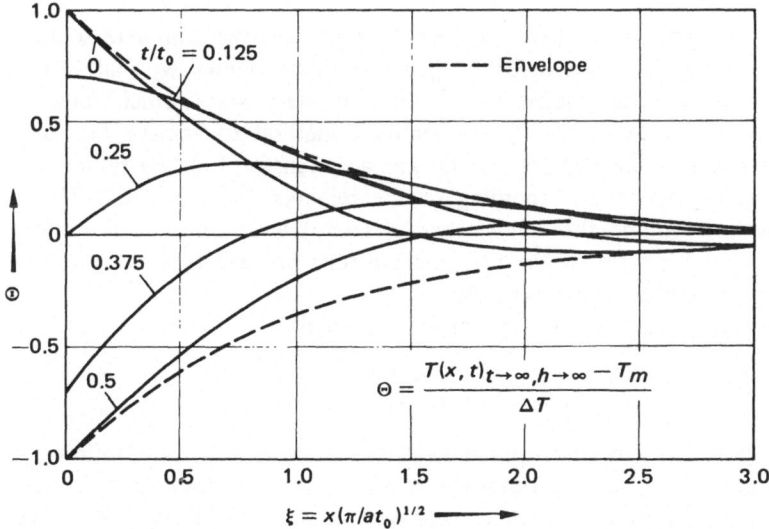

Figure 6.15 Temperature field in semi-infinite body when surface temperature varies periodically ($h \to \infty$).

The above ordinary differential equation for $\bar{T}$ has as its general solution the expression

$$\bar{T}(x, s) = C_1 \exp\left[-x\left(\frac{s}{a}\right)^{1/2}\right] + C_2 \exp\left[+x\left(\frac{s}{a}\right)^{1/2}\right]$$

It is easy to see that $C_2 = 0$, because $T(x, s)$ must remain bounded for $x \to \infty$. The boundary condition of the third kind allows us to calculate the remaining constant of integration, and we obtain

$$C_1 = \frac{(T_m/s) + s\,\Delta T/(s^2 + 4\pi^2/t_0^2)}{1 + (k/h)(s/a)^{1/2}}$$

We introduce the abbreviations

$$\left.\begin{aligned}
p &= \frac{2}{t_0} \\[2mm]
\xi &= x\left(\frac{\pi}{at_0}\right)^{1/2} \\[2mm]
\beta &= \frac{k}{h}\left(\frac{\pi}{at_0}\right)^{1/2}
\end{aligned}\right\} \tag{6.85}$$

and obtain the following Laplace transform of our solution

$$\bar{T}(x, s) = \frac{[(T_m/s) + s\,\Delta T/(s^2 + p^2)]\,\exp\,[-\xi(st_0/\pi)^{1/2}]}{1 + \beta(st_0/\pi)^{1/2}} \tag{6.86}$$

The rather complicated function (6.86) constitutes the complete solution to our problem in the transformed plane. In what follows, we shall restrict ourselves to the determination of the stationary state which must be expected to set in at $t \to \infty$. The simplest method to achieve this is to perform the limiting process on the Laplace transform (6.86) rather than on the actual solution. According to the remarks made in Sec. 6.10, we must now determine the singularities of (6.86) which possess the largest real value to the right of the real axis and then proceed to add the various terms which appear in the expansions.

We begin with the first term of (6.86), which is

$$\bar{T}_1(x, s) = \frac{T_m\,\exp\,[-\xi(st_0/\pi)^{1/2}]}{s[1 + \beta(st_0/\pi)^{1/2}]}$$

We recognize by inspection that at $s = 0$ there occurs a pole (an infinity) as well as a branch point. In order to determine the stationary state, it is not necessary to determine the whole expansion about $s = 0$, because it suffices to evaluate the limiting value for $s \to 0$. This gives

$$\bar{T}_1(x, 0) = \frac{T_m}{s}$$

and the first correspondence of Table 6.7 yields

$$T_1(x, \infty) = T_m \tag{6.87}$$

which is a constant.

The second term of function (6.86) possesses a branch point at $s = 0$ as well as simple poles at $s = +ip$ and $s = -ip$. The whole term vanishes for $s = 0$, which signifies that the singularity there makes no contribution to the periodic solution.

The sum of the remaining limiting values of the second term of function (6.86) must be calculated with the aid of the following relations

$$\left(+\frac{ipt_0}{\pi}\right)^{1/2} = 2^{-1/2}(1 + i)\left(\frac{pt_0}{\pi}\right)^{1/2} = 1 + i \qquad \text{in either case,}$$

the root with the

$$\left(-\frac{ipt_0}{\pi}\right)^{1/2} = 2^{-1/2}(1 - i)\left(\frac{pt_0}{\pi}\right)^{1/2} = 1 - i \qquad \text{largest real part}$$

and with the use of the identities

$$\frac{s}{s^2 + p^2} = \frac{1}{2(s - ip)} + \frac{1}{2(s + ip)}$$

After a rather lengthy calculation, we obtain, finally,

$$\bar{T}_2(x, s)_{s \to \pm p} = \frac{(\Delta T) \exp(-\xi)}{2(1 + 2\beta + 2\beta^2)}$$

$$\times \left[\frac{1 + \beta - i\beta}{s - ip} \exp(-i\xi) + \frac{1 + \beta + i\beta}{s + ip} \exp(+i\xi)\right] \tag{6.88}$$

The second correspondence in Table 6.7 leads us to an expression for the function in Eq. (6.88). This is obtained, first, in complex form. Employing Euler's formula

$$\exp(\pm i\phi) = \cos\phi \pm i\sin\phi$$

we can show, after another lengthy calculation, that this is equivalent to the real form

$$T_2(x, t)_{t \to \infty} = \frac{(\Delta T) \exp(-\xi)}{1 + 2\beta + 2\beta^2} [(1 + \beta)\cos(pt - \xi) + \beta\sin(pt - \xi)] \tag{6.89}$$

We introduce one more contraction, namely

$$\epsilon = \tan^{-1}\frac{\beta}{1 + \beta} \tag{6.90}$$

collect terms, and obtain

$$T_\infty(x, t) = T(x, t)_{t \to \infty}$$

$$= T_m + \frac{(\Delta T) \exp(-\xi)}{(1 + 2\beta + 2\beta^2)^{1/2}} \cos(pt - \epsilon - \xi) \qquad (6.91)$$

Equation (6.91) allows us immediately to calculate the solution for the boundary condition of the first kind simply by setting $h = \infty$. Equation (6.85) shows that then $\beta = 0$ and $\epsilon = 0$. Hence

$$T(x, t)_{\substack{t \to \infty \\ h \to \infty}} = T_m + (\Delta T) \exp(-\xi) \cos(pt - \xi) \qquad (6.92)$$

Both solutions show that the temperature field is represented by strongly damped oscillations of the same frequency as that of the forcing oscillation. The oscillation is also characterized by a phase shift, with respect to the forcing oscillation, which increases with x or ξ. The coefficient of heat transfer h affects the amplitude of the oscillation as well as the phase shift, separately from the effect of the depth x. For this reason, it suffices to plot the temperature field for $h = \infty$, as was done in Fig. 6.15, in order to form a picture of the temperature field even when $h \neq \infty$.

The equation of the envelope of the waves is determined by the application of the condition

$$\frac{\partial T_\infty(x, t)}{\partial t} = 0$$

to Eq. (6.91) which leads to

$$T_\infty^*(x, t) = T_m \pm \frac{(\Delta T) \exp(-\xi)}{(1 + 2\beta + 2\beta^2)^{1/2}} \qquad (6.93)$$

The temporal variation of the heat flux at the surface satisfies the relation

$$\dot{q}(0, t) = -k \left(\frac{\partial T_\infty}{\partial x} \right)_{x=0}$$

$$= \frac{2^{1/2} b \pi^{1/2} \Delta T}{t_0^{1/2} (1 + 2\beta + 2\beta^2)^{1/2}} \cos(pt - \epsilon + \tfrac{1}{4}\pi) \qquad (6.94)$$

with the penetration depth

$$b = (k\rho c_p)^{1/2}$$

The model adopted in the preceding example describes, at least approximately, important cases of periodic temperature fields that occur in nature and in engineering. We confine our remarks to the mention of the diurnal and annual temperature variations over the ground and of the high-frequency temperature variation in the walls of the cylinder of an internal combustion engine.

Example 6.9 The temperature of the gases inside the cylinder of a large diesel engine (two-stroke with $n = 120$ rev/min) performs an approximately harmonic fluctuation. The coefficient of heat transfer to the cylinder wall made of gray cast iron ($k = 50$ W/m K, $a = 14 \times 10^{-6}$ m^2/s) can be approximated by the mean value of $h = 1000$ W/m^2 K. For our purposes, we shall assume that the cylinder wall is very thick and not cooled. We are to determine the ratio ϕ_0 of the amplitude of the temperature fluctuation at the inner cylinder surface and in the gas, as well as the depth inside the wall at which this ratio has decreased to a value $\phi^* = 0.001$. The given data inserted into Eq. (6.85) yield

$$p = 2\pi n = 12.56 \text{ s}^{-1} \qquad \beta = 33.5 \qquad x = \xi \frac{m}{670}$$

According to Eq. (6.91), the amplitude ratio is

$$\phi = \frac{\exp{(-\xi)}}{(1 + 2\beta + 2\beta^2)^{1/2}}$$

Already at $x = \xi = 0$ the temperature amplitude has decreased to $\phi_0 = 2.08\%$, and the value $\phi^* = 0.001$ occurs at the very shallow depth of

$$x = m \times \ln{(\phi_0/\phi^*)}/670 = 4.53 \text{ mm}$$

The preceding numerical estimates demonstrate that the material strength of even low-speed internal combustion engines can be based on the assumption of a constant wall temperature.

Example 6.10 The stone sidewall of a cellar which separates it from the ground ($k = 0.6$ W/m K and $a = 0.2 \times 10^{-6}$ m^2/s) extends to a depth of about $x = 2$m under the surface. We wish to calculate the time interval Δt by which the temperature of the ground at depth x lags behind the annual temperature variation of the surroundings. The latter is assumed to vary harmonically with period $t_0 = 365$ d. We stipulate an annual, average coefficient of heat transfer at the surface of the earth of $h = 15$ W/m^2 K. Equation (6.85) yields $p = 0.0172$ d^{-1}; $\xi = 1.41$; $\beta = 0.028$, and Eq. (6.90) gives $\epsilon = 0.027$. We calculate the time lag Δt by setting the argument of the cosine in Eq. (6.91) equal to zero, which gives

$$\Delta t = \frac{\epsilon + \xi}{p} = 83.5 \text{ d}$$

that is, almost three months. This calculation explains why cellar walls feel cool and moist in summer and have a tendency to grow moss. By contrast, they feel dry in the winter.

SEVEN

THERMAL EXPLOSIONS

We say that a thermal explosion has occurred when a quantity of heat has been suddenly released in a volume whose extent can be neglected compared with the dimensions of the surroundings of interest. The heat may be released by a fast chemical or nuclear reaction, it may have its origin in electrical work (short circuit), or it may represent the enthalpy change in a rapid phase transformation, e.g., in fast condensation processes. The release of kinetic energy after a shot or meteorite impact can also serve as a source of heat. We begin with the discussion of a point source.

We assume that at instant $t = 0$ at a point $r = 0$ we have released an enthalpy H_0 [SI unit: Joule (J)] in an extended, homogeneous, and isotropic medium. We ask for the temperature field $T(r, t)$, where T is the excess temperature over the initial temperature, denoted by $T_0 = 0$. In accordance with the reasoning advanced earlier in Secs. 6.2 and 6.3, we make the tentative assumption that

$$T(r, t) = \exp(-\xi^2) \times f(t) \tag{7.1}$$

where the dimensionless coordinate is defined as

$$\xi = \frac{r}{(4at)^{1/2}}$$

Since no additional heat sources are postulated, it follows that the quantity of heat H_0 must be recovered at any instant in the form of an increase in the enthalpy of the surroundings. This means that the integral

$$\int_0^\infty 4\pi r^2 \rho c_p T \, dr = H_0 \tag{7.2}$$

is a constant and is independent of time. Note that we conduct our analysis here on the supposition that the medium is incompressible and that no deformation work is performed. Substituting Eq. (7.1) into (7.2), we find that

$$4\pi \rho c_p (4at)^{3/2} \int_0^\infty \xi^2 \exp(-\xi^2) \, d\xi \times f(t) = H_0 \tag{7.3}$$

In order for this expression to be independent of time, we must have

$$f(t) = \frac{C}{t^{3/2}}$$

Since, furthermore,

$$\int_0^\infty \xi^2 \exp(-\xi^2) \, d\xi = \tfrac{1}{4}\pi^{1/2}$$

we compute the constant C and find

$$C = \frac{H_0}{(4\pi a)^{3/2} \rho c_p}$$

Thus, we arrive at the solution

$$T(r, t) = \frac{H_0}{(4\pi at)^{3/2} \rho c_p} \exp\left(-\frac{r^2}{4at}\right) \tag{7.4}$$

This solution satisfies the following boundary conditions:

$$\text{at } t = 0 \text{ and } 0 < r < \infty \qquad \text{we have } T = 0$$

$$\text{at } t = 0 \text{ and } r = 0 \qquad \text{we have } T = \infty$$

$$\text{at } t = \infty \text{ and any } r \qquad \text{we have } T = 0$$

We are dealing here with a spherically symmetric problem. For this reason, our solution in Eq. (7.4) must satisfy the differential equation (6.9) with $n = 2$; this is the equation

$$\frac{\partial T}{\partial t} = a \frac{\partial^2 T}{\partial r^2} + \frac{2}{r} \frac{\partial T}{\partial r} \tag{7.5}$$

To supply the proof, we refer to Eq. (6.8) and re-arrange it in the form

$$\frac{\Theta}{r} = \frac{C}{4(at)^{3/2}} \exp\left(-\frac{r^2}{4at}\right) \tag{7.6}$$

We recall that this is a solution of Eq. (6.1), or of

$$\frac{\partial \Theta}{\partial t} = a \frac{\partial^2 \Theta}{\partial r^2} \tag{7.7}$$

We now proceed to calculate the derivatives that occur in Eq. (7.5) with respect to the function

$$T(r, t) = \frac{\Theta(r, t)}{r}$$

and so we are led back to Eq. (7.7). This proves that Eq. (7.4) is also a solution of Eq. (7.5).

The temperature distribution about a line source of length L (conceived as a segment of an infinite line) in which we have suddenly released enthalpy H_0 can be found in an analogous manner. Instead of Eq. (7.2), we employ the integral

$$\int_0^\infty 2\pi r L \rho c_p T \, dr = H_0 \tag{7.8}$$

The product hypothesis of Eq. (7.1) leads to

$$8\pi \rho c_p \, at \int_0^\infty \xi \exp\left(-\xi^2\right) d\xi \times f(t) = \frac{H_0}{L} \tag{7.9}$$

This expression becomes independent of time for $f(t) = C/t$. Since

$$\int_0^\infty \xi \exp\left(-\xi^2\right) d\xi = \frac{1}{2}$$

we calculate

$$C = \frac{H_0/L}{4\pi a \rho c_p}$$

and arrive at the solution

$$T(r, t) = \frac{H_0/L}{4\pi at \rho c_p} \exp\left(-\frac{r^2}{4at}\right) \tag{7.10}$$

Inserting this function into Eq. (6.9) with $n = 1$, we can satisfy ourselves that Eq. (7.10) represents the required solution for the problem of cylindrical symmetry.

In the case of a heat source in the shape of a flat plate of area $2A$ (being a portion of an infinite plane) that suddenly releases enthalpy H_0, we would obtain

$$T(r, t) = \frac{H_0/A}{(4\pi at)^{1/2}\rho c_p} \exp\left(-\frac{r^2}{4at}\right) \tag{7.11}$$

Here, $\frac{1}{2}H_0$ represents the heat that moves into a half-space and r is the distance measured from the plane of the source. Equation (7.11) corresponds to the fundamental solution of Eq. (6.7).

The three solutions Eqs. (7.4), (7.10), and (7.11) can be put in the following, more symmetric, dimensionless form:

$$\frac{\pi^{1/2} T\rho c_p r}{H_0/A} = \xi \exp\left(-\xi^2\right) \qquad \text{(flat plate)} \tag{7.12}$$

$$\frac{\pi T\rho c_p r^2}{H_0/L} = \xi^2 \exp\left(-\xi^2\right) \qquad \text{(cylinder)} \tag{7.13}$$

$$\frac{\pi^{3/2} T\rho c_p r^3}{H_0} = \xi^3 \exp\left(-\xi^2\right) \qquad \text{(sphere)} \tag{7.14}$$

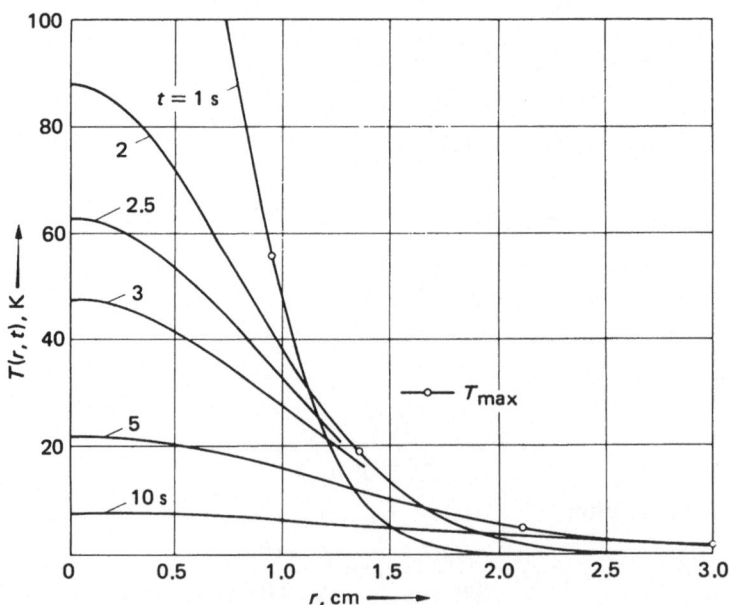

Figure 7.1 Temperature distribution $T(r, t)$ about an instantaneous point-source of heat. See Example 7.1. T_{max} is the highest temperature achieved at distance r.

In all cases $\xi = r/(4at)^{1/2}$. The three functions which appear on the right-hand side in the preceding solutions have been graphed in Fig. 6.1.

Example 7.1 A thick steel plate, which may be modeled as a semi-infinite body, is characterized by $a = 15 \times 10^{-6}$ m^2/s and $\rho c_p = 3900$ kJ/m^3 K. The plate is hit by a bullet of mass $m = 10$ g moving with a velocity $w = 500$ m/s. We make the crude assumption that the bullet does not penetrate the plate, but deforms and is brought to a stop at the surface of the plate. This is equivalent to saying that the bullet transfers its total kinetic energy $\frac{1}{2}mw^2 = 1.25$ kJ to the plate by conduction. With these simplifications, we can calculate the temperature distribution in the plate with the aid of Eq. (7.4) if we further neglect the finite space occupied by the deformed bullet. Here, the energy imparted to the semi-infinite space is $\frac{1}{2}H_0 = 1.25$ kJ. With the numerical data provided, Eq. (7.4) assumes the form

$$T(r,\ t) = 248 t^{-1.5} \exp(-\xi^2)Ks^{1.5} \qquad (7.15)$$

with $\xi = r^2/(4at)$. The temperature distribution is represented by a characteristic Gaussian bell-curve. At every point the temperature distribution passes through a maximum that occurs at instant $t_m = r^2/6a$, Fig. 7.1.

EIGHT

CONTINUOUSLY OPERATING HEAT SOURCES

We consider a homogeneous and isotropic body of infinite extension. At $r = 0$ there is placed a continuously operating heat source that transfers heat at a rate $\dot{Q}(t)$ (SI unit: Watt, W). The quantity of heat transferred from time $t = \tau$ to $t = \tau + d\tau$ is $\dot{Q}(t)\, d\tau$; according to Eq. (7.4), this heat generates a temperature field. Since the Laplace equation is linear, the temperature fields generated over time are superimposed on each other, and we can write the solution for a continuously operating heat source in the form

$$T(r, t) = \frac{1}{(4\pi a)^{3/2} \rho c_p} \int_0^t \dot{Q}(t) \exp\left(-\frac{r^2}{4a(t - \tau)}\right) \times \frac{d\tau}{(t - \tau)^{3/2}}$$

$$(8.1)$$

The substitution

$$\phi = \frac{r}{[4a(t - \tau)]^{1/2}} \qquad \text{and hence} \qquad d\phi = \frac{r\, d\tau}{4a^{1/2}(t - \tau)^{3/2}}$$

allows us to find that

$$T(r, t) = \frac{1}{2\pi^{3/2} kr} \int_{r/(4at)^{1/2}}^{\infty} \dot{Q}(\tau) \exp\left(-\phi^2\right) d\phi \qquad (8.2)$$

In the case of a source of constant output, this integrates to

$$T(r, t) = \frac{\dot{Q}}{4\pi kr} \ \text{erfc} \ \frac{r}{(4at)^{1/2}} \tag{8.3}$$

where the complementary error function, erfc ξ, has been defined in Eq. (6.23). Equation (8.3) describes the temperature field as it develops around a point source of constant rate from the instant when it has been switched on. For $t \to \infty$ we obtain the steady-state heat flow

$$\dot{Q} = 4\pi rkT \tag{8.4}$$

which agrees with Eq. (3.10) if we interpret T as denoting the excess temperature at radius r over that of the surroundings.

The temperature field formed around a continously operating cylindrical heat source emitting at a rate $\dot{Q}(\tau)/L$, referred to length L, is calculated by integrating Eq. (7.10). This yields

$$T(r, t) = \frac{1}{4\pi k} \int_0^t \frac{\dot{Q}(\tau)}{L} \ \exp\left[-\frac{r^2}{4a(t-\tau)}\right] \times \frac{d\tau}{t-\tau} \tag{8.5}$$

We use the substitution

$$\chi = \frac{r^2}{4a(t-\tau)} \quad \text{and hence} \quad \frac{d\tau}{t-\tau} = \frac{4a(t-\tau)\,d\chi}{r^2} = \frac{d\chi}{\chi}$$

We can integrate Eq. (8.5) for a constant rate $\dot{Q}/L = \dot{Q}_L$ to produce

$$T(r, t) = \frac{\dot{Q}_L}{4\pi k} \int_{t^2/4at}^\infty \frac{\exp(-\chi)}{\chi} \ d\chi = -\frac{\dot{Q}_L}{4\pi k} \ Ei\left(-\frac{r^2}{4at}\right) \tag{8.6}$$

Here we have introduced the exponential integral $Ei(x)$ defined by the relation

$$-Ei(-x) = \int_x^\infty \frac{\exp(-u)}{u} \ du$$

which can be found tabulated in Appendix G2.[†]

The temperature field generated by a continuously operating source of strength $\dot{Q}(t)$ in the form of a plane sheet of area $2A$ is calculated in an analogous manner, that is, by the integration of Eq. (7.11); in this case:

$$T(r, t) = \frac{1}{(4\pi a)^{1/2}\rho c_p} \int_0^t \frac{\dot{Q}(\tau)}{A} \ \exp\left[-\frac{r^2}{4a(t-\tau)}\right] \frac{d\tau}{(t-\tau)^{1/2}} \tag{8.7}$$

[†]Often, the notation $Ei_1 x$ is used instead of $-Ei(-x)$.

Once again we employ the substitutions

$$\phi = \frac{r}{[4a(t-\tau)]^{1/2}} \qquad \text{and hence} \qquad d\phi = \frac{r \, d\tau}{4a^{1/2}(t-\tau)^{3/2}}$$

and integrate Eq. (8.7) for $\dot{Q}$ equal to a constant. Thus

$$T(r, t) = \frac{r\dot{Q}}{4Ak\pi^{1/2}} \int_{r/(4at)^{1/2}}^{\infty} \frac{\exp(-\phi^2)}{\phi^2} \, d\phi$$

$$= \frac{r\dot{Q}}{k2A} \left[\frac{(4at)^{1/2}}{r\pi^{1/2}} \exp\left(-\frac{r^2}{4at}\right) - \text{erfc} \, \frac{r}{(4at)^{1/2}} \right] \qquad (8.8)$$

Example 8.1 A current-carrying wire is embedded in a synthetic material (for example, in polyvinyl chloride) whose thermal properties are: ρ, c_p, and k. At $t = 0$ heat is released at a constant rate $\dot{Q}_L$ per unit length. The temperature in the material at a radius r, where r is much larger than the wire radius, can be measured as a function of time t with the aid of a very thin thermocouple, which is also embedded in the synthetic material. In order to compare the results of such a measurement with the theory derived earlier for the case of a cylindrical heat source, that is with a theory that can only apply by way of approximation, we prepare an auxiliary diagram. This plots the reduced temperature

$$\theta = \frac{4\pi k T(r, t)}{\dot{Q}_L}$$

vs. the reduced time

$$\text{Fo} = \frac{at}{r^2} = \frac{kt}{\rho c_p r^2}$$

The graph in Fig. 8.1 is such a diagram; it was plotted with the aid of Table G.2 in the Appendix, and represents

$$\theta = -Ei\left(\frac{-1}{4\text{Fo}}\right)$$

We suppose that for our material we can find numerical values only for ρ and c_p, but do not know its thermal conductivity, k. Every pair of values T_i and t_i can be used to calculate θ_i and Fo_i with an estimated value for k. Such points can be entered in Fig. 8.1. The physically correct value of k can be said to have been found when the measured points closely follow the curve in the graph.

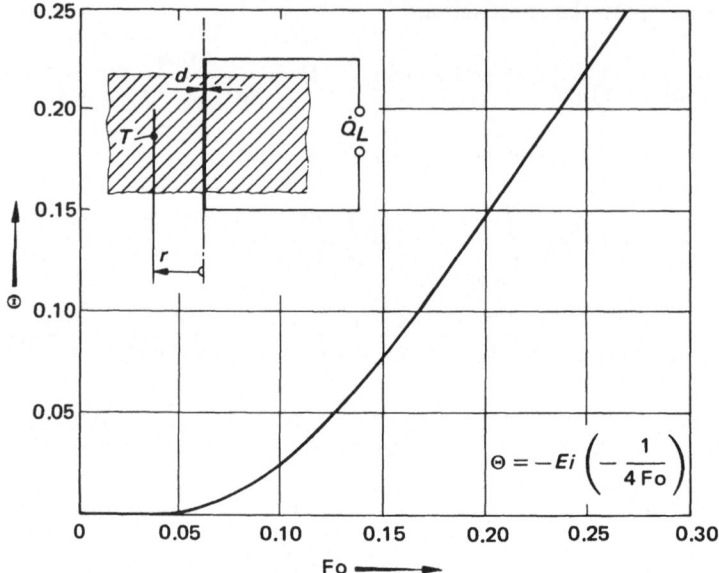

Figure 8.1 Variation of temperature at distance r from a cylindrical source of constant output.

It[†] may be worth noting that the exponential integral has a convenient series expansion (see [11.1], p. 262) valid for small values of the argument $x = r^2/4at$, that is, for large times and sufficiently small values of the radius r; its form is

$$Ei(-x) = \gamma + \ln x - x + \cdots$$

Here $\gamma = 0.5772 \ldots$ is Euler's constant. With the aid of this expansion, Eq. (8.6) can be simplified to the expression

$$T(r, t) = -\frac{\dot{Q}_L}{4\pi k}\left(\gamma + \ln\frac{r^2}{4a} - \ln t - \frac{r^2}{4at} + \cdots\right)$$

$$\approx \frac{\dot{Q}_L}{4\pi k}\left(\ln t - \gamma - \ln\frac{r^2}{4a}\right) \tag{a}$$

This equation is of the form

$$T(r, t) \approx \frac{A}{k}(\ln t - B) \tag{b}$$

[†]Paragraph added by translator with the authors' consent.

where A and B are constants during a measurement. It follows that a plot of $T(r, t)$ vs. ln t is a straight line whose slope is inversely proportional to the thermal conductivity, the factor A being known. The intercept $-AB/k$ can be ignored in such a plot, because its value does not affect the slope of the straight line of T vs. ln t. If we operate in the regime where the approximation (a), and hence (b), is valid, there is no need to know ρ and c_p to measure k whose value is, moreover, calculated without the need to resort to trial and error.

The preceding principle has been successfully used to measure the thermal conductivity of fluids with great accuracy.

NINE

MOVING SOURCES

In Chapter 7 we determined the temperature fields $T(r, t)$ that formed about point sources, line sources, and plane sources of heat. In all cases, H_0 denoted the total amount of heat released suddenly at $r = 0$ (thermal explosion).

A range of important technological processes can be analyzed approximately with the aid of models that assume that a heat source of given extent moves through a medium at rest or that a stationary heat source is placed in a medium that moves through it with a constant velocity. Examples of problems of this kind are found in machining or welding processes. In the succeeding analysis, we prefer the latter representation. Accordingly, we fix the origin of a suitably oriented system of rectangular coordinates x, y, z at the tip of the cutting tool or of the welding electrode. We then assume that the piece being machined or welded moves across with a velocity w in the negative x direction, which thus becomes the preferred direction.

Following the same procedure as in Chapter 8, we assume a constant heat rate $\dot{Q}$ of the source. The heat released during time $d\tau$ in the interval $0 \leqslant \tau \leqslant t$ is $dH_0 = \dot{Q}\,d\tau$. The description of the temperature field formed about a moving point source to which we have attached the origin of a rectangular coordinate system is based on Eq. (7.4) to which we apply two modifications. First, we replace the duration t of the influence of the heat source by the interval $t - \tau$. Secondly, the radius vector r, which in Eq.

(7.4) is measured from each instantaneous position of the thermal explosion, must now be expressed with respect to its instantaneous position in the reference system x, y, z, which is at rest. This relation is

$$r^2 = [x - w(t - \tau)]^2 + y^2 + z^2$$

Further, we replace H_0 in Eq. (7.4) by dH_0, and derive the expression

$$dT = \frac{\dot{Q}\,d\tau}{[4\pi a(t - \tau)]^{3/2}\rho c_p}\,\exp\left\{-\frac{[x - w(t - \tau)]^2 + y^2 + z^2}{4a(t - \tau)}\right\}$$

The required temperature field $T(x, y, z, t)$ is obtained by the superposition of all infinitesimal temperature fields which result from a chain of thermal explosions of strength $dH_0 = \dot{Q}\,d\tau$ in the time interval $0 \leqslant \tau \leqslant t$. This is done by integration in which we employ the abbreviation $R^2 = x^2 + y^2 + z^2$ and the substitutions

$$\phi = \frac{R}{[4a(t - \tau)]^{1/2}} \qquad \text{and hence} \qquad d\phi = \frac{R\,d\tau}{4a^{1/2}(t - \tau)^{3/2}}$$

The result is

$$T(x, y, z, t) = \frac{\dot{Q}}{2Rk\pi^{3/2}}\left[\exp\left(\frac{wx}{2a}\right)\right] \times \int_{R/(4at)^{1/2}}^{\infty} \exp\left(-\phi^2 - \frac{w^2 R^2}{16a^2\phi^2}\right) d\phi$$

$$(9.1)$$

The temperature field reaches a steady state for $t \to \infty$. In this case the lower limit of the integral vanishes, and Eq. (9.1) leads us to the simple expression

$$T(x, y, z) = \frac{\dot{Q}}{4\pi kR}\,\exp\left[-\frac{w(R - x)}{2a}\right] \qquad (9.2)$$

It is convenient to introduce the dimensionless quantities

$$\Theta = \frac{4\pi kaT}{w\dot{Q}}$$

$$\xi = \frac{wx}{a}$$

$$\eta = \frac{wy}{a}$$

$$\zeta = \frac{wz}{a}$$

$$\gamma^2 = \eta^2 + \zeta^2$$

$$(9.3)$$

and to write the solution as

$$\theta(\xi, \eta, \zeta) = \frac{\exp\{-0.5\,[(\xi^2 + \gamma^2)^{1/2} - \xi]\}}{(\xi^2 + \gamma^2)^{1/2}} \tag{9.4}$$

The temperature field represented by Eq. (9.4) is seen plotted in Fig. 9.1 in the plane ξ, η at $\zeta = 0$. Three isotherms $\theta = $ const are presented.

Starting with Eq. (7.10) for the steady-state temperature field formed about a cylindrical source of length L and strength $\dot{Q}_L = \dot{Q}/L$ placed along the y axis, we can employ an analogous sequence of steps. We would thus calculate

$$T(x, z) = \frac{\dot{Q}_L}{2\pi k}\left[\exp\left(\frac{wx}{2a}\right)\right] K_0\left[\frac{w}{2a}\,(x^2 + z^2)^{1/2}\right] \tag{9.5}$$

The symbol K_0 denotes the modified Bessel function of the second kind listed in Appendix Table G.3.

Introducing

$$\Theta_L = \frac{2\pi k T}{\dot{Q}_L} \tag{9.6}$$

we derive the dimensionless form of the solution:

$$\Theta_L(\xi, \zeta) = [\exp\left(\tfrac{1}{2}\xi\right)]\,K_0\left[\tfrac{1}{2}(\xi^2 + \zeta^2)^{1/2}\right] \tag{9.7}$$

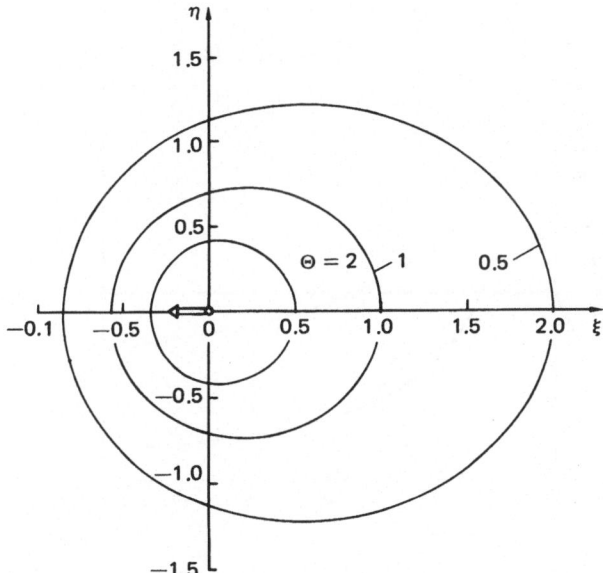

Figure 9.1 Temperature field about a point source that moves to the left.

As a last example in this series, we solve the problem of a two-dimensional source placed in the plane $y = z = 0$ whose strength over the total area $2A$ is $\dot{Q}$. Then

$$
T(x) = \begin{cases} \dfrac{\dot{Q}}{A\rho c_p w} & \text{for } x \geq 0 \\[2ex] \dfrac{\dot{Q}}{A\rho c_p w}\ \exp\left(\dfrac{wx}{a}\right) & \text{for } x < 0 \end{cases}
\tag{9.8}
$$

We put

$$
\Theta_A = \frac{A\rho c_p wT}{\dot{Q}}
\tag{9.9}
$$

and rewrite Eq. (9.8) in the form

$$
\Theta_A(\xi) = \begin{cases} 1 & \text{for } \xi \geq 0 \\ \exp \xi & \text{for } \xi < 0 \end{cases}
\tag{9.10}
$$

The reduced temperatures θ, θ_L, θ_A for a point source, a line source, and a plane source, respectively, have been plotted in terms of the coordinate ξ in Fig. 9.2. The curves represent Eqs. (9.4), (9.7), and (9.10).

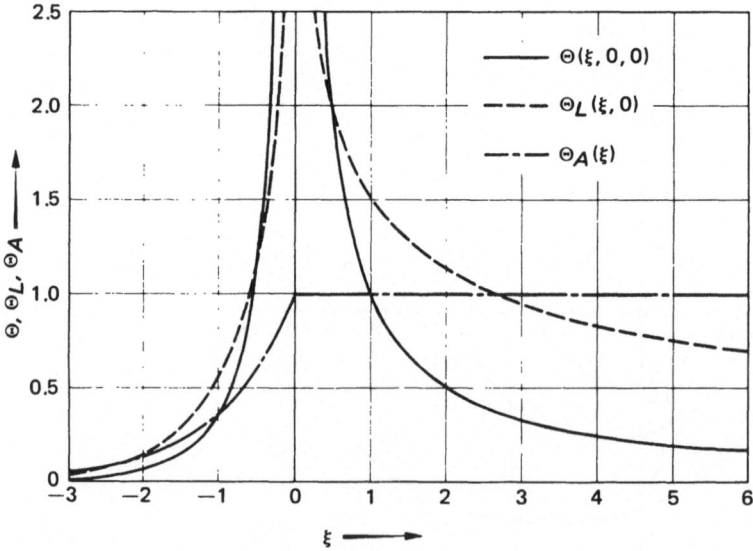

Figure 9.2 Temperature distribution along the ξ axis for a point source, line source, and plane source that move to the left.

In Eq. (9.4) we have put $\gamma = 0$, and in Eq. (9.7) we assumed $\zeta = 0$. Thus, in effect, the curves represent the functions $\Theta(\xi, 0, 0)$, $\theta_L(\xi, 0)$, and $\theta_A(\xi)$.

Example 9.1 A seam has been welded onto a thick block of steel. The welding was executed with the aid of an electrode of 3.25 mm in diameter. The current was one of $I = 140$ A and the arc voltage (dc current) was $U = 25$ V; thus the total electric energy expended was $Q_E = 3.5$ kW. According to [9.1] it is possible to assume, by way of a crude estimate, that a fraction $\epsilon = 0.8$ of the evolved heat penetrated the steel block. The same reference further indicates that during the arc-welding process it is necessary to perform the calculations using the properties of steel at 400°C. In this case, these are

$$\rho = 7730 \text{ kg/m}^3 \quad k = 45 \text{ W/m K} \quad c = 600 \text{ J/kg K} \quad a = 96 \times 10^{-6} \text{ m}^2/\text{s}$$

The melting temperature is $T_m \approx 1470$°C, and the welding speed was 3.3 mm/s during which a weld 10 mm wide was produced. The initial temperature of the steel block was $T_0 \approx 20$°C. It is required to calculate the width of the welding seam and the width of the zone which underwent a change of its grain structure. (Such structural changes occur in steels for temperatures $T > 400$°C.)

The arc is conceived as a moving point source that is displaced along the surface of a semi-infinite body. This allows us to use Eqs. (9.2)–(9.4) with $\dot{Q} = 2\epsilon\dot{Q}_E$. The temperature field on the surface of the block (the ξ-η plane) in the moving coordinate system is described by Eq. (9.4) with $\gamma = \eta$ (see also Fig. 9.1). To calculate the isotherm, we put

$$(\xi^2 + \eta^2)^{1/2} = \mu$$

and transform Eq. (9.4) to read

$$\xi = \mu + 2 \ln (\mu\Theta) \tag{9.11}$$

The corresponding value of the abscissa ξ is calculated from Eq. (9.11) for a given value of the temperature Θ on the isotherm, and for a prescribed radius μ drawn from the origin. The corresponding value of the abscissa η is calculated with

$$\eta = (\mu^2 - \xi^2)^{1/2}$$

(The data in Fig. 9.1 have been calculated according to this equation.)

The preceding calculational model does not take into account the finite extent of the arc or the existence of a melted zone; therefore, we must expect that the discrepancies between theory and reality near the weld are quite large. In a real situation, both phenomena cause heat to flow in a direction normal to that of the moving electrode. Consequently, the isotherms shown in Fig. 9.1 must move away from each other in a direction normal to the ξ axis. We suppose that the melting temperature T_m is reached exactly at the edges of the welding seam. The reduced temperature is then

$$\Theta = \frac{4\pi ka(T_m - T_0)}{2w\epsilon\dot{Q}_E} = 0.426$$

The isotherm $\Theta = 0.5$ in Fig. 9.1 conveys an idea of the shape of the moving zone inside which $T > T_m$ prevails. In principle, half the width of the seam—5 mm in the experiment—is determined by the largest half-width of the egg-shaped zone inside the reduced isotherm $\Theta = 0.426$. We calculate

$$\eta_{max} = \left(\frac{w}{a}\right) y_{max}$$

from Eq. (9.11) by iteration. To accomplish this, we may prescribe different values for the radius μ estimated with the aid of Fig. 9.1. The result is:

$$\eta_{max} = 1.37 \qquad \text{at } \xi = 0.67$$

$$y_{max} = 4 \text{ mm}$$

This result is in satisfactory agreement with the value found experimentally ($y_{max} = 5$ mm), because we realize that the preceding calculation must yield values of the width that are too small. To complete our task, we now calculate the extent of the zone of structural change. Assuming $T = 400°C - T_0 = 380$ K, we would calculate on the basis of our earlier data that $\Theta = 0.112$, $\eta_{max} = 3.23$ at $\xi = 3.1$, and $y_{max} \approx 10$ mm. It is seen that the zone of structural change extends approximately over two seam widths.

TEN

NONSTEADY HEAT CONDUCTION IN SEVERAL DIMENSIONS

In Secs. 6.3 and 6.6, we derived the functions that describe the temperature distributions in a semi-infinite body, the flat plate, the cylinder, and the sphere. The assumption was made that the temperature of the body at instant $t = 0$ was uniform and equal to T_c. We made the further assumption that the temperature of the fluid surrounding the system was kept constant at T_∞ for $t > 0$. In the general case, we prescribed a heat transfer coefficient h that constitutes a boundary condition of the third kind. As far as the three geometrical shapes mentioned explicitly above are concerned, we recognize that the preceding solutions are valid under the assumption that the bodies extend to infinity in a direction perpendicular to the direction of the heat flux. However, it is possible to utilize the same solutions for a wider class of body shapes. Let us imagine that we have traced a contour on the surface of the flat plate. The resulting area may be confined within a closed curve or it may extend to infinity in some directions. Let us further imagine that we create a cut along the contour curve at right angles to the surface and into the depth of the body, making sure that the surfaces of the cut are adiabatic. It is clear that temperature distributions inside the bodies resulting from such cuts continue to be described by the same solutions.

In the case of a cylinder or sphere, we make the cut (more precisely, an infinite sequence of infinitely small drill holes) through an arbitrary contour on the surface, at right angles to the surface, and ensure that the cut passes through the center of the sphere or at right angles through the axis of the cylinder. The corresponding solutions from Sec. 6.6 retain their validity if the cuts are made along adiabatic surfaces. We now proceed to discuss a special set of bodies created as a result of such adiabatic cuts. This will be the set of bodies which can arise as the result of a mutual interpenetration of two or three of the fundamental shapes: semi-infinite body, plane plate, infinitely long cylinder and sphere. We shall impose the additional requirement that the heat fluxes that exist in the basic forms in the case of nonsteady, one-dimensional heat conduction must be mutually perpendicular in the combined shapes. This requirement allows us to create two combinations with cylindrical symmetry, but none with the sphere! Figures 10.1a through i and Table 10.1 describe the bodies that we have in mind.

We now assume that the body exists at a uniform temperature T_c and exchanges heat with a fluid of temperature $T_\infty = 0$ also through the surfaces of the cuts. In such circumstances, there will form a nonsteady, three-

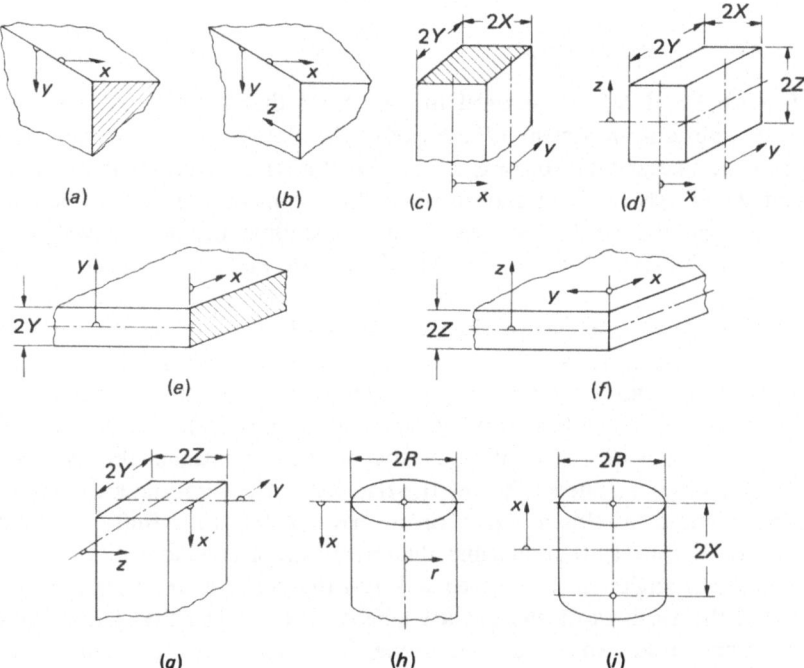

Figure 10.1a–i Geometric shapes for which descriptions can be obtained. They are characterized by heat conduction in several dimensions. Adiabatic surfaces are shown hatched.

Table 10.1 Geometric shapes for which descriptions can be obtained. They are characterized by heat conduction in several dimensions

S: semi-infinite body   F: flat plate   C: infinitely long cylinder

Figure 10.1 (part)	Combination of basic shapes	Description of resulting model
a	S/S	Rectangular edge of a large body
b	S/S/S	Rectangular corner of a large body
c	F/F	Long prismatic bar with rectangular cross-section
d	F/F/F	Slab
e	S/F	Flat plate cut off at one end
f	S/S/F	Flat plate cut by two mutually perpendicular planes
g	S/F/F	Prismatic bar of rectangular cross-section cut off at one end
h	S/C	Cylinder cut off at one end
i	F/C	Cylinder of finite length

dimensional temperature field that can be simply described with the aid of the elementary one-dimensional solutions already known to us. The temperature variation in the interior or on the surface of the new models can be calculated with the aid of products of elementary solutions. The latter are contained in Eq. (6.38) or (6.44) for the plate, Eq. (6.39) or (6.45) for the cylinder, and Eq. (6.47) for the semi-infinite body [6.2]. In what follows, we shall use the symbols X, Y, Z for the half-thicknesses of flat plates or slabs and R for the radius of a cylinder; x, y, z, r will denote the corresponding internal coordinates. The latter symbols will also be used for the coordinates inside a semi-infinite body.

We verify the preceding assertion with the example of the rectangular slab represented in Fig. 10.1c. The temperature field in such a body must satisfy the two-dimensional form of the Fourier equation (1.4), namely

$$\frac{\partial T}{\partial t} = a \left(\frac{\partial^2 T}{\partial x^2} + \frac{\partial^2 T}{\partial y^2} \right)$$

We assert that the product of the basic solutions for two flat plates of thickness $2X$ and $2Y$ satisfies the above differential equation. We assume that the basic solutions are given formally by the expressions

$$T(x, t) = \Phi \left(\frac{x}{X}, \frac{at}{X^2} \right) \quad \text{and} \quad T(y, t) = \Psi \left(\frac{y}{Y}, \frac{at}{Y^2} \right)$$

The product solution is then

$$T(x, y, t) = \Phi \Psi$$

We denote partial derivatives by subscripts, $viz.$, Φ_t, Φ_{xx}, Ψ_t, and Ψ_{yy}. With these expressions substituted into the differential equation, we find that

$$\Phi_t \Psi + \Phi \Psi_t = a(\Phi_{xx} \Psi + \Phi \Psi_{yy})$$

or that

$$\Psi(\Phi_t - a\Phi_{xx}) + \Phi(\Psi_t - a\Psi_{yy}) = 0$$

The basic solutions satisfy the relations

$$\Phi_t = a\Phi_{xx} \qquad \text{and} \qquad \Psi_t = a\Psi_{yy}$$

and thus automatically satisfy the above. It is remarkable that each of the one-dimensional equations may contain different Fourier and Biot numbers; this means that it is possible to prescribe different coefficients of heat transfer on each face. In the case of a flat plate it is necessary to recognize two exceptions: either the Biot number must have the same value on the two parallel faces or else it must vanish on one face (adiabatic wall). In the latter case, the Biot and Fourier numbers, Bi and Fo, must be formed with the full plate thickness. Using similar stratagems, it is possible to construct subordinate geometries to those given in Figs. 10.1a through i and Table 10.1.

We shall now illustrate the preceding method of calculation with the aid of an example.

Example 10.1 We are given a cylinder made of chrome-nickel steel of diameter $2R = 0.6$ m and height $2X = 0.5$ m. The thermal properties are: $k = 14.5$ W/m K, and $a = 3.85 \times 10^{-6}$ m^2/s. We are to find the reduced temperatures in the center and on the external surfaces of the cylinder at $t = 1.5$ h $= 5400$ s after a jump in the temperature of the surrounding fluid. The coefficient of heat transfer on the flat faces is $h = 40$ W/m^2 K, whereas that on the cylindrical surface is $h_R = 60$ W/m^2 K. The preceding values define the following dimensionless quantities:

$$\text{Fo} = \frac{at}{X^2} = 0.333 \qquad \text{Bi} = \frac{hX}{k} = 0.690$$

$$\text{Fo}_R = \frac{at}{R^2} = 0.231 \qquad \text{Bi}_R = \frac{h_R R}{k} = 1.242$$

It has been explained in Sec. 6.7 that for Fo $\geqslant 0.24$ (plate) and Fo$_R \geqslant 0.21$ (cylinder) it is possible to use the approximate solutions of Eqs. (6.41) and (6.42). Thus, according to the data in Table 6.2, the reduced excess temperatures T_m/T_c and T_w/T_c in the middle and on the surface of a flat plate of thickness $2X$ (added subscript P) become

$$\frac{T_{mP}}{T_c} = 0.905 \qquad \frac{T_{wP}}{T_c} = 0.665$$

Similarly, by interpolation in Table 6.3, we find the corresponding numerical values for a cylinder of radius R (added subscript C):

$$\frac{T_{mC}}{T_c} = 0.812 \qquad \frac{T_{wC}}{T_c} = 0.476$$

Forming products based on the preceding four numerical values, we calculate at the center

$$\Theta_1 = \frac{T_{mP}}{T_c} \frac{T_{mC}}{T_c} = 0.735$$

in the center of the flat faces

$$\Theta_2 = \frac{T_{wP}}{T_c} \frac{T_{mC}}{T_c} = 0.540$$

and at the rim of the flat faces

$$\Theta_3 = \frac{T_{wP}}{T_c} \frac{T_{wC}}{T_c} = 0.316$$

Finally, at the center line of the cylindrical surface, we calculate

$$\Theta_4 = \frac{T_{mP}}{T_c} \frac{T_{wC}}{T_c} = 0.430$$

The product solutions retain their validity at any point inside the body, and are not restricted to the approximate solutions. In the preceding example we could as well have employed Eqs. (6.44) and (6.45) or the complete solutions Eqs. (6.38) and (6.39), because for the plate-cylinder combination we must have, quite generally,

$$\frac{T(x, r, t)}{T_c} = \frac{T(x, t)}{T_c} \frac{T(r, t)}{T_c} = \frac{T_P}{T_c} \frac{T_C}{T_c} \tag{10.1}$$

Regarding the shapes in Fig. 10.1c, d, and i that do not involve a semi-infinite body, it is possible to calculate the heat loss Q. This is easy to obtain from the product solutions for the temperature field. In the case of a cylinder of finite length, we calculate that

$$Q_c = \pi R^2 2X \rho c_p T_c$$

is the total amount of heat that can be transferred. Thus, according to Eq.

(10.1), we must then have

$$Q = Q_c \left(1 - \frac{1}{Q_c} \rho c_p T_c \int_V \frac{T}{T_c} \, dV \right)$$

$$= Q_c \left(1 - \frac{1}{\pi R^2 2X} \int_{-X}^{+X} \frac{T_P}{T_c} \, dx \int_0^R 2\pi r \frac{T_C}{T_c} \, dr \right) \quad (10.2)$$

In an analogous way, we obtain for the flat plate (supplementary subscript P):

$$Q_P = Q_{cP} \left(1 - \frac{1}{2X} \int_{-X}^{+X} \frac{T_p}{T_c} \, dx \right) \quad (10.3)$$

and for the infinitely long cylinder (added subscript C)

$$\Theta_C = \Theta_{cC} \left(1 - \frac{1}{\pi R^2} \int_0^R 2\pi r \frac{T_C}{T_c} \, dr \right) \quad (10.4)$$

Substituting Eqs. (10.3) and (10.4) into (10.2) we obtain the modified product form

$$\frac{Q}{Q_c} = 1 - \left[1 - \left(\frac{Q}{Q_c} \right)_P \right] \left[1 - \left(\frac{Q}{Q_c} \right)_C \right] \quad (10.5)$$

Employing the approximate relation in Eq. (6.43) and the numerical values from Tables 6.2 and 6.3, we calculate with the data of this example:

$$\left(\frac{Q}{Q_c} \right)_P = 0.175 \qquad \left(\frac{Q}{Q_c} \right)_C = 0.363 \qquad \frac{Q}{Q_c} = 0.475$$

Further details concerning problems of this general nature can be found in [6.2].

ELEVEN

NONSTEADY HEAT CONDUCTION WITH PHASE TRANSITION

Pure substances, eutectic alloys, and mixtures change their state of aggregation or their allotropic form at fixed transition temperatures that are characteristic of each material. During such a phase transition, the system liberates or stores the associated latent heat (enthalpy of the transformation). Noneutectic alloys and mixtures perform the transition within a fixed temperature interval and are included in the succeeding considerations only when mentioned explicitly.

In a wide class of conduction processes in nature and technology, the system becomes divided into two or more regions by the transition isotherms. These phase boundaries traverse the body when the process is not stationary. Across the phase boundaries, the properties, k, ρ, c_p, etc., change discontinuously.

Due to the fact that the phase boundaries may move through the body and, above all, owing to the appearance of a nonlinear boundary condition at the phase boundary, the mathematical analysis of such processes becomes much more difficult.

In what follows, we shall analyze the solidification of a liquid, and assume, by way of approximation, that the thermal properties of the two regions on the two sides of the phase boundary are different, but do not change with temperature on either side. We exclude convection in the liquid phase, so that we face a pure conduction problem in both phases.

The penetration of the frost line into the ground or in water-containing foods (meat, fruit, vegetables) and the freezing of stagnant pools of water provide us with concrete examples of such problems. In the last case, we observe the appearance of a stable stratification without convection which is due to the anomalous properties of water in the temperature interval $0°C \leqslant T \leqslant 4°C$.

In the year 1891, J. Stefan solved a problem of this class concerning the formation of ice in the polar seas. He analyzed a semi-infinite body of constant surface temperature (boundary condition of the first kind) assuming that at $t = 0$ the liquid phase is at the melting temperature T_m. The case when the initial temperature over the whole expanse of the liquid exceeds the melting temperature T_m was treated in his lectures by F. Neumann since the 1860s. He did not actually publish the solution until the year 1912. Since that time, no essentially new solutions of physical importance have been discovered, either for boundary conditions of the second or third kind, or for bodies of other shapes.

11.1 NEUMANN'S EXACT SOLUTION

Let a semi-infinite body have a uniform temperature $T_2(x_2, 0) = T_{II}$ at instant $t = 0$. The subscripts 2 and II will henceforth be reserved for the liquid phase. We assume that $T_{II} > T_m$. The free surface of the body will be kept at temperature $T_1(0, t) = T_I$ at all times $t > 0$. The subscripts 1, I will characterize the solid phase. It is assumed that $T_I < T_m$; see Fig. 11.1.

In cases where the densities ρ_1 and ρ_2 of the liquid and solid phases differ from each other, at least one of the two phases must move with respect to an observer assumed to be at rest. In other words, not only does the phase boundary move through the system, but masses must also actually be displaced. In a natural lake that freezes, the ice sheet formed at instant t must move away from the bottom, because $\rho_1 < \rho_2$ for H_2O. Simultaneously, the thickness of the ice layer grows at the underside. By contrast, if water freezes in a container which can only be cooled from the bottom, the growing layer of ice will lift the remaining layer of water upward. The density anomaly of water gives rise to inner convective currents, and for this reason the latter case cannot be solved with the aid of the following scheme.

For ease of mathematical description, it is convenient to anchor the space coordinate x_1 of the solid phase at the surface of the solidified layer. For the same reason, it is convenient to link the space coordinate x_2 of the liquid phase with the plane which coincided with the surface of the body at $t = 0$. If $s_1(t)$ and $s_2(t)$ denote the coordinates of the phase boundary at

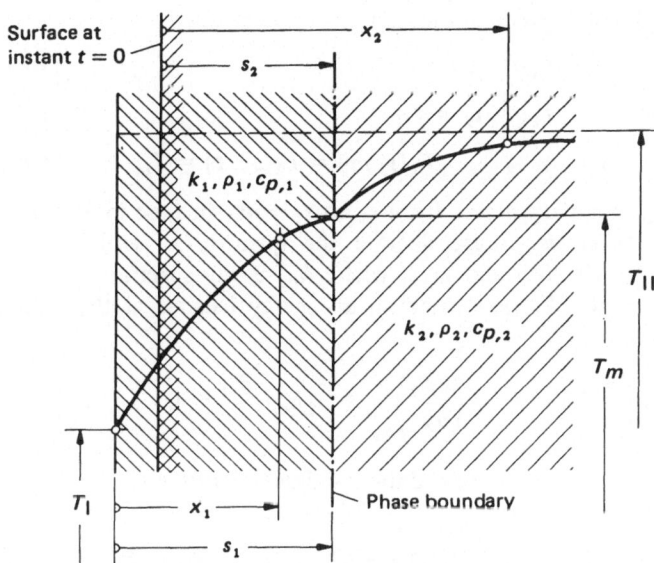

Figure 11.1 Penetration of the solidification boundary in a semi-infinite medium.

instant t, then the law of mass conservation dictates that we must have

$$\rho_1\, ds_1 = \rho_2\, ds_2 \qquad \text{or} \qquad s_1\rho_1 = s_2\rho_2 \qquad (11.1)$$

whereas the coordinates x_1' and x_2' of a point located in the interior of the solid or liquid phase must obey the transformation law

$$x_1' - x_2' = s_1 - s_2 \qquad (11.2)$$

The convention regarding the origin of coordinates x_1 and x_2 specified earlier imposes, as its consequence, the property that neither phase is shifted with respect to the coordinate system associated with it, except for the boundary of the region. It follows, further, that a Fourier differential equation of the form of Eq. (1.2) must be valid in each of the domains. Thus,

$$\frac{\partial T_1}{\partial t} = a_1\, \frac{\partial^2 T_1}{\partial x_1^2} \qquad \text{in the interval } 0 \leqslant x_1 \leqslant s_1 \qquad (11.3)$$

$$\frac{\partial T_2}{\partial t} = a_2\, \frac{\partial T_2}{\partial x_2^2} \qquad \text{in the interval } s_2 \leqslant x_2 \leqslant \infty \qquad (11.4)$$

The initial conditions are

$$t = 0: s_1 = 0 \qquad s_2 = 0 \qquad T_2(x_2, 0) = T_{\mathrm{II}} \qquad (11.5)$$

Three of the four boundary conditions can be stated at once:

$$x_1 = 0: \qquad T_1(0, t) = T_I \tag{11.6}$$

$$x_2 \to \infty: \qquad T_2(\infty, t) = T_{II} \tag{11.7}$$

$$x_1 = s_1 \text{ or } x_2 = s_2: \qquad T_1(s_1, t) = T_2(s_2, t) = T_m \tag{11.8}$$

When the freezing boundary moves from s_1 to $s_1 + ds_1$ in time interval dt, the layer of thickness ds_1 must liberate the quantity of enthalpy of transformation $\rho_1 h_m \, ds_1$ per unit area, where h_m is the specific enthalpy of melting (latent heat). We now establish an energy balance at the phase boundary, and write

$$-k_1 \left(\frac{\partial T_1}{\partial x_1}\right)_{s_1} + k_2 \left(\frac{\partial T_2}{\partial x_2}\right)_{s_2} + \rho_1 h_m \frac{ds_1}{dt} = 0 \tag{11.9}$$

It is now useful to introduce dimensionless quantities for ease of manipulation. These are

$$\Theta_1 = \frac{T_1 - T_I}{T_m - T_I} \qquad \Theta_2 = \frac{T_2 - T_m}{T_{II} - T_m} \qquad \Theta = \frac{T_{II} - T_m}{T_m - T_I} \tag{11.10}$$

$$\text{Fo}_{x1} = \frac{a_1 t}{x_1^2} \qquad \text{Fo}_{x2} = \frac{a_2 t}{x_2^2} \tag{11.11}$$

$$\alpha = \frac{a_2}{a_1} = \frac{k_2 c_{p1} \rho_1}{k_1 c_{p2} \rho_2} \tag{11.12}$$

$$\beta = \frac{b_2}{b_1} = \left(\frac{k_2 c_{p2} \rho_2}{k_1 c_{p1} \rho_1}\right)^{1/2} \tag{11.13}$$

$$\gamma = \frac{\rho_2}{\rho_1} \tag{11.14}$$

$$\text{Sn}^\dagger = \frac{h_m}{c_{p1}(T_m - T_I)} \tag{11.15}$$

To solve the problem, we tentatively postulate that the solutions are of the form

$$\Theta_1 = A \text{ erf } [\tfrac{1}{2}(\text{Fo}_{x1})^{1/2}] \tag{11.16}$$

$$\Theta_2 = 1 - B \text{ erfc } [\tfrac{1}{2}(\text{Fo}_{x2})^{1/2}] \tag{11.17}$$

It can be verified that these forms satisfy the differential equations (11.3)

---

†Translator's note: Even though no general agreement seems to exist, we propose to refer to this dimensionless group as the Stefan number Sn.

and (11.4), the initial conditions, Eq. (11.5), and the boundary conditions, Eqs. (11.6) and (11.7), due attention being paid to the definitions, Eqs. (11.10) and (11.11). We substitute Eqs. (11.16) and (11.17) into the boundary condition, Eq. (11.8) and obtain two equations for the three unknowns: A, B, s_1, and time t, if we also make use of Eqs. (11.1), (11.12), and (11.14). Hence

$$\frac{1}{A} = \text{erf } \frac{s_1}{2(a_1 t)^{1/2}}$$

$$\frac{1}{B} = \text{erfc } \frac{s_1}{2\gamma(\alpha a_1 t)^{1/2}}$$

The preceding two relations can be satisfied for all values of time only on condition that the phase coordinate s_1 is proportional to $t^{1/2}$. This condition translates to the nondimensional form

$$\delta = \frac{s_1}{2(a_1 t)^{1/2}} \qquad \text{or} \qquad \text{Fo}_{s1} = \frac{a_1 t}{s_1^2} \tag{11.18}$$

Equation (11.18) turns the constants A and B into functions of the characteristic number δ, that is of the special Fourier number $\text{Fo}_{s1} = 1/(4\delta^2)$. (In the literature of the subject some authors seem to favor the nondimensional numbers δ or $\delta' = 2\delta$ over Fo_{s1}.) In this manner, we are led to

$$A = \frac{1}{\text{erf } \delta} \qquad \text{and} \qquad B = \frac{1}{\text{erfc } (\delta/\gamma\alpha^{1/2})} \tag{11.19}$$

The energy balance, Eq. (11.9), at the phase boundary leads first, after some calculation, to a transcendental equation which determines δ and, secondly, the Stefan or phase-transition number Sn, which is characteristic of nonsteady conduction problems accompanied by a phase transition. The Stefan number has been defined in Eq. (11.5) in anticipation of its occurrence here. The equation is

$$\delta\pi^{1/2}\,\text{Sn} = \frac{1}{\exp \delta^2 \times \text{erf } \delta} - \frac{\beta\Theta}{\exp (\delta^2/\gamma^2\alpha) \times \text{erfc } [\delta/(\gamma\alpha^{1/2})]} \tag{11.20}$$

Equation (11.20) makes it possible to calculate by iteration δ as a function of the three characteristic dimensionless groups: $\text{Sn}, \beta\Theta$, and $\gamma\alpha^{1/2}$. We now quote without derivation the limiting solution of Eq. (11.20):

$$\delta \ll 1: \quad \delta \approx \frac{1}{2\epsilon\pi^{1/2}} \, [\beta^2\Theta^2 + 2\pi\epsilon)^{1/2} - \beta\Theta] \tag{11.21}$$

where the contraction

$$\epsilon = Sn + \frac{1}{3} + \frac{2\beta\Theta}{\pi\gamma\alpha^{1/2}} \tag{11.22}$$

has been introduced.

Figure 11.2 depicts δ as a function of Sn with $\beta\Theta$ and $\gamma\alpha^{1/2}$ as parameters.

Substituting the expressions in Eq. (11.19) and performing the transformation of Eq. (11.2) to the coordinate anchored in the surface of the ice, we derive the following two equations for the reduced temperatures from Eqs. (11.16) and (11.17):

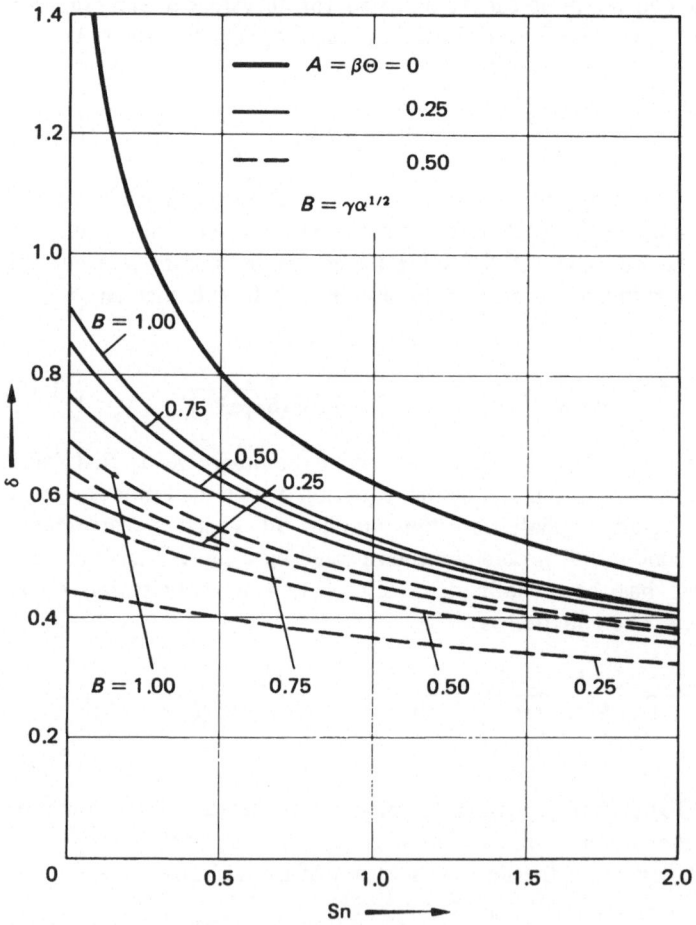

Figure 11.2 Parametric representation of the function $\delta = \delta(Sn, \beta\Theta, \gamma\alpha^{1/2})$.

$$\Theta_1 = \frac{\text{erf}\,[\tfrac{1}{2}(\text{Fo}_{x1})^{1/2}]}{\text{erf}\,\delta} \qquad (11.23)$$

$$\Theta_2 = 1 - \frac{\text{erfc}\,\{\alpha^{-1/2}\,[1/(2\,\text{Fo}_{x1}^{1/2}) + (1-\gamma)\delta/\gamma]\}}{\text{erfc}\,(\delta/\gamma\alpha^{1/2})} \qquad (11.24)$$

The second term in the argument of the function (11.24) accounts for the movement of the liquid phase with respect to the coordinate x_1 which is fixed in the solid phase.

Reference [11.1] catalogs all analytic solutions of the Stefan-Neumann problem known so far, as well as their variants. This reference contains also the solution for the case of a finite density jump at the phase boundary, which is incorrectly presented in many reference books.

Example 11.1 Thick filets of fish are quick-frozen on board a fishing vessel by being suddenly immersed into Freon 12, which boils at a temperature of $-30°C$. The thickness of the filets is 3 cm and their initial temperature is $T_{II} = T_m = 0°C$. We wish to determine the time t^* after the lapse of which the two frost-fronts will meet at the center, that is at $s_2^* = 1.5$ cm. We assume that the filets of fish have the shape of slabs. We, further, wish to determine the freezing rate ds_2/dt at t^*.

The present example is studied with the intention of providing us with only very rough estimates. For this reason, we assign to the fish the properties of water and ice (see Table 11.1). We ignore special effects, such as the freezing-point depression and partial freezing.

Thermophysical properties:

$$\rho_1(-15°C) = 919 \text{ kg/m}^3 \qquad \rho_2(0°C) = 1000 \text{ kg/m}^3$$

$$c_{p1}(-15°C) = 1980 \text{ J/kg K} \qquad k_1(-15°C) = 2.4 \text{ W/m K}$$

$$h_m = 333 \text{ kJ/kg}$$

Since $T_2 = T_m = 0°C$, we note that the Stefan-Neumann solution is valid for a plate of finite thickness, and particularly, up to the time t^* at which the two fronts meet. The Stefan number from Eq. (11.15) has the value

$$\text{Sn} = \frac{h_m}{c_{p1}(T_m - T_I)} = 5.61$$

Both the approximate equation (11.21) and the exact equation (11.20) yield the value $\delta = 0.290$ for $\text{Sn} = 5.61$ with $\Theta = 0$. Equations (11.1) and (11.18) yield the freezing time

$$t^* = \left(\frac{s_2\rho_2}{\delta\rho_1}\right)^2 \frac{1}{4a_1} = \left(\frac{0.015 \times 1000}{0.290 \times 919}\right)^2 \times \frac{10^6}{5.28} \text{ s} = 10 \text{ min}$$

Table 11.1 Thermophysical properties of liquid water and ice at $p = 1$ bar

	Property value								
Liquid									
t, °C	0	10	20	30	40	50	60	80	100
ρ_2, kg/m³	1000	1000	998	996	992	988	983	972	958
c_{p2}, J/kg K	4220	4190	4180	4180	4180	4180	4180	4200	4220
k_2, W/mK	0.552	0.578	0.598	0.614	0.628	0.641	0.652	0.669	0.682
Solid (ice)									
t, °C	0	−20	−40	−60	−80	−100	−120	−150	−180
ρ_1, kg/m³	917	920	923	925	927	929	931	933	934
c_{p1}, J/kg K	2040	1950	1810	1650	1520	1390	1250	1040	820
k_1, W/mK	2.25	2.45	2.7	3.0	3.5	4.0	4.6	5.7	7.2

At this instant, the freezing velocity is

$$\left(\frac{ds_2}{dt}\right)_{t^*} = \frac{\rho_2}{\rho_1}\left(\frac{ds_1}{dt}\right)_{t^*} = \frac{\rho_2 \delta_1 a_1^{1/2}}{\rho_1 t^{*1/2}} = 5.3 \text{ cm/h}$$

The further customary freezing of the filet to a temperature of $T = -20$°C (in the center) occurs without phase transition and can be analyzed, for example, with the aid of the Binder-Schmidt method described in Sec. 6.8.

11.2 THE QUASI-STEADY APPROXIMATIONS

The Stefan number

$$\text{Sn} = \frac{h_m}{c_{p1}(T_m - T_I)}$$

is the ratio of the total "latent heat" (enthalpy of transformation)

$$H_m = m h_m$$

to total "sensible heat" (enthalpy of temperature change)

$$H = m c_{p1}(T_m - T_I)$$

Both quantities must be extracted from a body of mass m on freezing when the process starts with freezing the liquid at the melting temperature T_m and ends by further cooling the solid uniformly to temperature $T_I < T_m$.

During nonsteady phase transitions, as seen from Fig. 11.1, the solid phase exists at an average temperature $\bar{T}$ (in the case of a plane domain), where $\bar{T} \geqslant \frac{1}{2}(T_I + T_m)$. Accordingly, it is necessary to extract merely the "sensible heat"

$$H^* = mc_{p1}(T_m - \bar{T}) \leqslant \frac{1}{2} mc_{p1}(T_m - T_1) = \frac{1}{2}H$$

Thus, the ratio of the enthalpy quantities actually extracted up to time t is

$$\frac{H_m}{H^*} \geqslant \frac{2h_m}{c_{p1}(T_m - T_1)} = 2 \text{ Sn}$$

In the example discussed in the preceding section, we calculated that $\text{Sn} = 5.61$. This meant that in that example H_m was at least 11 times larger than H^*. If we now prescribe a boundary condition of the third kind and make T_I equal to the mean cooling temperature, we find that for the wall temperature $T_w > T_I$ and the mean temperature $\bar{T} \geqslant \frac{1}{2}(T_w + T_m)$, the ratio H_m/H^* increases even further. The process of solidification is more and more dominated by the enthalpy of transformation H_m.

According to the preceding argument, it is permissible to neglect H^* with respect to H_m in the limiting case when $\text{Sn} \to \infty$. This means that in the solid zone of thickness s_1 we need to take into account only the resistance due to heat conduction, as is the case with rigorously stationary problems. In the case of large, but finite values $\text{Sn} \gg 1$, it is possible to derive from this model explicit formulas; they are known as quasi-steady approximations.

The diagrams of Fig. 11.3a and b schematically represent the distribution of temperature during the process of solidification on the inner and outer surface of a hollow cylinder—a case which frequently occurs in practical applications.

A coolant of temperature $T_I < T_m$ extracts heat from the inner or

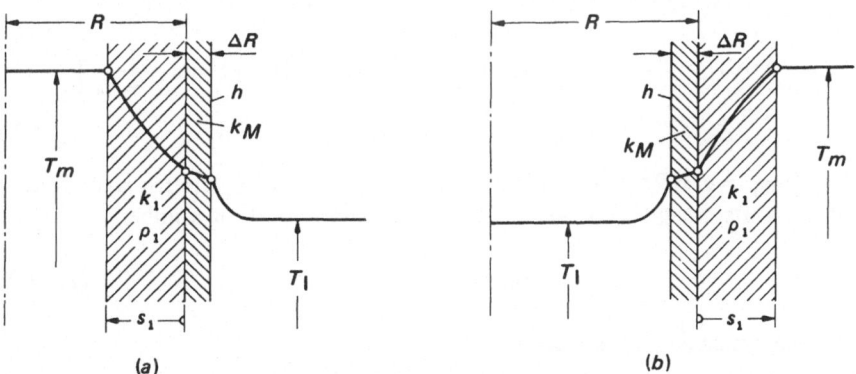

(a)   (b)

Figure 11.3 Temperature distribution during solidification on the (a) inner surface and (b) outer surface of a hollow cylinder. (Boundary condition of the third kind.)

outer surface of a tube (inner diameter R, wall thickness ΔR, thermal conductivity of tube material k_M). The film coefficient that operates on the cooled side is h. In practice, we always encounter cases when $\Delta R/R \ll 1$. For this reason, we can determine the overall coefficient of heat transfer as for a flat plate, i.e., in accordance with Eq. (3.13) of Sec. 3.2. Thus

$$U = \frac{1}{1/h + \Delta R/k_M} \qquad (11.25)$$

We proceed now to give a detailed analysis of the case of exterior cooling with the initial condition that the tube is completely filled with a fluid at the melting temperature T_m at instant $t = 0$. This means that we assume $T_{II} = T_m$. Applying the modified Péclet equation (3.15) for the cylinder derived in Sec. 3.2, we can establish the following energy balance at instant t and up to the location $s_1(t)$ of the phase boundary. This reads

$$\frac{2\pi(T_m - T_I)}{1/[U(R + \Delta R)] + (1/k_1)\ln[R/(R - s_1)]} = 2\pi\rho_1 h_m(R - s_1)\frac{ds_1}{dt} \qquad (11.26)$$

Integration of this separable differential equation subject to the initial condition that $s_1(0) = 0$ and with the contractions

$$Bi_e = \frac{U(R + \Delta R)}{k_1} \qquad (11.27)$$

$$\sigma_1 = \frac{s_1}{R} \qquad (11.28)$$

leads to the explicit solution

$$Fo_{R1} = \frac{Sn}{2}\left[\sigma_1(2 - \sigma_1)\left(\frac{1}{2} + \frac{1}{Bi_e}\right) - (1 - \sigma_1)^2\ln\frac{1}{1 - \sigma_1}\right] \qquad (11.29)$$

for $Fo_{R1} = a_1 t/R^2$. When the tube is cooled internally, the ice layer forms on the outside and we must work with the modified Biot number

$$Bi_i = \frac{U(R - \Delta R)}{k_1} \qquad (11.30)$$

An analogous derivation yields

$$Fo_{R1} = \frac{Sn}{2}\left[(1 + \sigma_1)^2\ln(1 + \sigma_1) - \sigma_1(2 + \sigma_1)\left(\frac{1}{2} - \frac{1}{Bi_i}\right)\right] \qquad (11.31)$$

In order to describe the process of solidification on the inner or outer side of a hollow sphere, we must introduce the modified Biot numbers

$$\text{Bi}_e = \frac{U(R + \Delta R)^2}{k_1 R} \tag{11.32}$$

and
$$\text{Bi}_i = \frac{U(R - \Delta R)^2}{k_1 R} \tag{11.33}$$

The quasi-steady approximations for the inner and outer surfaces, respectively, are

$$\text{Fo}_{R1} = \text{Sn}\left\{\frac{1}{2}\left[1 - (1 - \sigma_1)^2\right] - \frac{1}{3}\left(1 - \frac{1}{\text{Bi}_e}\right)\left[1 - (1 - \sigma_1)^3\right]\right\} \tag{11.34}$$

$$\text{Fo}_{R1} = \text{Sn}\left\{\frac{1}{3}\left(\frac{1}{\text{Bi}_i} + 1\right)\left[(1 + \sigma_1)^3 - 1\right] - \frac{1}{2}\left[(1 + \sigma_1)^2 - 1\right]\right\} \tag{11.35}$$

The approximate calculation of a solidification problem on a flat plate of wall thickness Δx is very simple, and there is even no need to make a distinction between the inner and outer surface. With the definition

$$U = \frac{1}{1/h + \Delta x/k_M} \tag{11.36}$$

we can write the phase boundary condition in the form

$$\frac{T_m - T_I}{1/U + s_1/k_1} = \rho_1 h_m \frac{ds_1}{dt} \tag{11.37}$$

Integration yields

$$\text{Fo}_{s_1} = \frac{a_1 t}{s_1^2} = \text{Sn}\left(\frac{1}{2} + \frac{1}{\text{Bi}_{s1}}\right) \tag{11.38}$$

where Fo_{s1} is calculated from Eq. (11.18), and

$$\text{Bi}_{s1} = \frac{Us_1}{k_1} \tag{11.39}$$

is formed with the time-varying coordinate s_1 of the phase boundary. Inserting the limiting value $\text{Bi}_{s1} \to \infty$ into the solution (11.38) of the plane problem, and taking into account the definition in Eq. (11.18), we derive an approximate expression (subscript A) for the characteristic number δ_A:

$$\delta_A = (2\text{Sn})^{-1/2} \tag{11.40}$$

In the special case when $T_{\mathrm{II}} = T_m$, that is, $\Theta = 0$, the exact solution follows from Eq. (11.20), and is

$$\delta \pi^{1/2} \, \mathrm{Sn} = \frac{1}{(\exp \delta^2)\,(\mathrm{erf}\,\delta)}$$

Figure 11.4 shows the relative error

$$\Delta \delta = \frac{\delta_A - \delta}{\delta}$$

plotted in terms of Sn. The arrows point to values of the Stefan number Sn which correspond to water and various metal melts. In the case of the metals, the diagram assumes $T_{\mathrm{I}} = 20°\mathrm{C}$, whereas for water the assumption $T_{\mathrm{I}} = -30°\mathrm{C}$ has been made. The value of c_{p1} is that which occurs at the melting point T_m in all cases. It is seen that the approximate solutions yield good results only for $\mathrm{Sn} > 2$, that is for water or substances which contain a high percentage of water. It is recalled from Eq. (11.18) that

$$t = \frac{s_1^2}{4a_1 \delta^2}$$

On the other hand, Fig. 11.4 shows that for steels $\Delta\delta \approx 42\%$, or, equivalently, that $\delta_A/\delta \approx 1.42$. It follows that the time needed to solidify

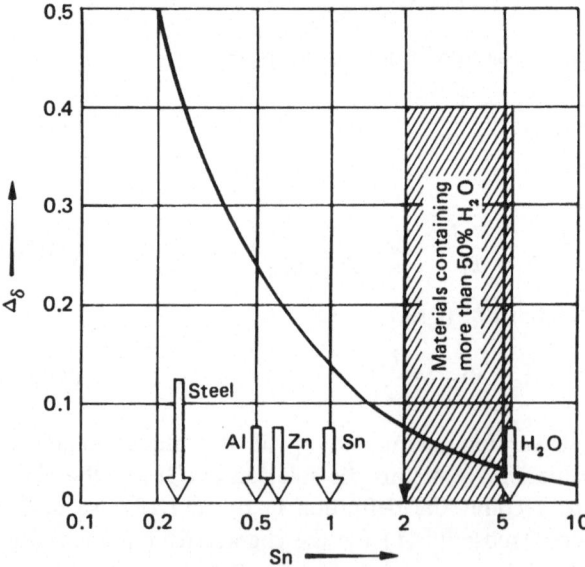

Figure 11.4 Error contained in the quasi-static approximation (plane geometry).

a layer of thickness s_1, and calculated with the aid of the approximate solution, is only $t_A \approx 0.5t$, that is, one half of the exact time t. For equal values of Sn, the errors in the calculation of solidification times of cylindrical or spherical bodies are even larger. In all cases when the asymptotic approximations yield values of inadequate accuracy, it is possible to fall back on the results of more accurate calculations contained in [11.2].

To conclude these considerations, it is necessary to draw the reader's attention to the influence of the change in density on solidification. The approximate solutions (11.29) and (11.34) for the inner volume of a cylinder or sphere, respectively, are valid without additional errors only when, with $\rho_2 > \rho_1$, as in water, the liquid can displace itself freely. If the opposite is true, that is for $\rho_2/\rho_1 < 1$, as is the case with most other liquids, the solution remains valid only if the liquid is replenished. If this is not the case (sphere!), the solid may burst or develop cavities.

Example 11.2 A substance containing water and characterized by h_m, k_1, ρ_1, c_{p1} exists at its melting point $(T_{II} = T_m)$ and in the following configurations:

(a) Plate of thickness $2X$ (cooling on both sides)
(b) Cylinder of radius $R_e = X$ (external cooling)
(c) Sphere—analogous to (b)
(d) Cylindrical shell of inner radius $R_i = X$, and outer radius $R_e = 2X$ (cooling only on the inside)
(e) Spherical shell—analogous to (d)

In all cases, we stipulate constant values of the cooling-medium temperature, T_I and of the film coefficient h. If we neglect the thickness of the container wall $(\Delta R = \Delta X = 0)$, we may put $U = h$ and

Table 11.2 Comparison of freezing times

Case	Equation	$\tau^*(Bi) = Fo_{x_1}^* / Sn$	τ^*	
			$Bi = 1$	$Bi \to \infty$
(a)	(11.38)	$1/2 + 1/Bi$	1.5	0.50
(b)	(11.29)	$(1/2 + 1/Bi)/2$	0.75	0.25
(c)	(11.34)	$1/2 - (1 - 1/Bi)/3$	0.50	0.167
(d)	(11.31)	$[4 \ln 2 - 3(1/2 - 1/Bi)]/2$	2.136	0.636
(e)	(11.35)	$7(1 + 1/Bi)/3 - 3/2$	3.167	0.833

$s_1 = X$ (complete freezing of body), and conclude that all Biot numbers are equal:

$$Bi_{s1} = Bi_f = Bi_e = Bi$$

We are to calculate the reduced freezing time

$$\tau^* = \frac{Fo^*_{X1}}{Sn} = \frac{t^* k_1 (T_m - T_I)}{\rho_1 h_m X^2}$$

for $Bi = 1$ and $Bi \rightarrow \infty$. The results are listed in Table 11.2.

REFERENCES

2.1. Sommerfeld, A. and Bethe, H.: *Elektronentheorie der Metalle.* Berlin, Göttingen, Heidelberg: Springer, 1967.

Justi, E.: *Leitungsmechanismus und Energieumwandlung in Festkörpern,* 2d ed. Göttingen, 1965.

Erdmann, J. C.: *Wärmeleitung in Metallen.* Berlin, Göttingen, Heidelberg: Springer, 1969.

Klemens, P. G.: Thermal Conductivity of Solids at Low Temperatures. In: *Handbuch der Physik,* S. Flügge, ed., vol. 14, pp. 198–291. Berlin, Göttingen, Heidelberg: Springer, 1956.

2.2. Cezairliyan, A. and Touloukian, Y. S.: Correlation and Prediction of Thermal Conductivity of Metals. In: *Advances in Thermophysical Properties at Extreme Temperatures and Pressures.* S. Gratch, ed., pp. 301–313. ASME: New York, 1965.

2.3. Bungardt, K. and Spyra, W.: *Arch. Eisenhüttenwes.,* vol. 36, pp. 257–267, 1965.

5.1. Kirchhoff, G.: *Vorlesungen über die Theorie der Wärme,* p. 13, Leipzig, 1894.

5.2. Höckel, J., Saur, G., and Borchers, H.: *J. Nucl. Mater.,* vol. 33, pp. 225–241, 1969.

5.3. Betz, A.: *Konforme Abbildung,* 2d ed., Berlin, Göttingen, Heidelberg: Springer, 1964.

Rothe, R., Ollendorf, F., and Polhausen, K.: *Funktionentheorie und ihre Anwendung in der Technik.* Berlin, 1931.

Knopp, K.: *Funktionentheorie,* 2 vols., 11th ed. de Gruyter, Berlin, 1965.

5.4. Hahne, E. and Schällig, R.: Formfaktoren der Wärmeleitung für Anordnungen mit isothermen Rippen. *Wärme-Stoffübertrag.,* vol. 5, pp. 39–46, 1972.

5.5. Hahne, E. and Grigull, U.: Formafaktor und Formwiderstand der stationären mehrdimensionalen Wärmeleitung. *Int. J. Heat Mass Transfer,* vol. 18, pp. 751–767, 1975.

6.1. Grigull, U.: *Temperaturausgleich in Einfachen Körpern.* Berlin, Göttingen, Heidelberg: Springer, 1964.

6.2. Grigull, U., Bach, J., and Sandner, H.: Näherungslösungen der Nichtstationären Wärmeleitung. *Forsch. Ingenieurwes.,* vol. 32, no. 1, pp. 11–18, 1966.

157

6.3. Schuh, H.: *Differenzenverfahren zum Berechnen von Temperatur-Ausgleichs-vorgängen* ... VDI Forschungsh. 459. Düsseldorf: VDI-Verlag, 1957.

6.4. Doetsch, G.: *Anleitung zum praktischen Gebrauch der Laplace-Transformation und der Z-Transformation.* 3d ed. München, Wien: R. Oldenbourg, 1967.

6.5. Baehr, H. D.: Die Lösung nichtstationärer Wärmeleitungsprobleme mit Hilfe der Laplace-Transformation. *Forsch. Ingenieurwes.*, vol. 21, pp. 33–40, 1955.

9.1. Rykalin, N. N.: *Berechnung der Wärmevorgänge beim Schweißen.* Berlin: VEB Verlag Technik, 1957.

11.1. Carslaw, H. S. and Jaeger, J. C.: *Conduction of Heat in Solids.* Oxford: Oxford University Press, 1959.

11.2. Stephan, K. and Holzknecht, B.: Die Asymptotischen Vorgänge des Erstarrens. *Int. J. Heat Transfer*, vol. 19, pp. 597–602, 1976. Wärmeleitung beim Erstarren geometrisch einfacher Körper. *Wärme-Stoffübertrag*, vol. 7, pp. 200–207, 1974.

INTERNATIONAL SYSTEM OF UNITS

The International System of Units (Système International d'Unités, or SI) is constructed with the aid of seven base units, the secondary SI units, and the derived units which are formed from them with the aid of a factor of unity, that is of consistent derived units. There exists one and only one SI unit for every physical quantity.

Table A.1 SI base units

Physical quantity	Unit name	Symbol
Length	Meter	m
Mass	Kilogram	kg
Time	Second	s
Electric current	Ampere	A
Thermodynamic temperature	Kelvin	K
Amount of substance	Mole	mol
Luminous intensity	Candela	cd

Table A.2 Derived SI units that are given separate names

Physical quantity	Unit name	Symbol	Structure
Frequency	Hertz	Hz	s^{-1}
Force	Newton	N	$kg\ m/s^2$
Pressure, mechanical stress	Pascal	Pa	$kg/m\ s^2 = N/m^2$
Energy, work, heat	Joule	J	$kg\ m^2/s^2 = Nm$
Power, rate of heat transfer	Watt	W	$kg\ m^2/s^3 = J/s$
Quantity of electricity or electric charge	Coulomb	C	$A\ s$
Electric potential difference, voltage, or electromotive power	Volt	V	$kg\ m^2/A\ s^3 = W/A$
Electric resistance	Ohm	Ω	$kg\ m^2/A^2\ s^3 = V/A$
Electric conductance	Siemens	S	$A^2\ s^3/kg\ m^2 = A/V$
Electric capacity	Farad	F	$A^2\ s^4/kg\ m^2 = A\ s/V$
Magnetic flux	Weber	Wb	$kg\ m^2/A\ s^2$
Magnetic flux density	Tesla	T	$kg/A\ s^2 = Wb/m^2$
Inductance	Henry	H	$kg\ m^2/A^2\ s^2 = Wb/A$
Luminous flux	Lumen	lm	$cd\ sr$
Illumination	Lux	lx	$cd\ sr/m^2$
Activity	Becquerel	Bq	s^{-1}
Energy dose	Gray	Gy	$m^2/s^2 = J/kg$
Plane angle (angle)	Radian	rad	m/m
Solid angle	Steradian	sr	m^2/m^2

A special name for the kelvin, when temperatures are indicated on the Celsius scale, is the degree Celsius (denoted °C). The kelvin is also a unit for temperature differences and temperature intervals.

Table A.3 Prefixes and symbols for powers of 10

Power	Prefix	Symbol	Power	Prefix	Symbol
10^{18}	Exa	E	10^{-1}	Deci	d
10^{15}	Peta	P	10^{-2}	Centi	c
10^{12}	Tera	T	10^{-3}	Milli	m
10^9	Giga	G	10^{-6}	Micro	μ
10^6	Mega	M	10^{-9}	Nano	n
10^3	Kilo	k	10^{-12}	Pico	p
10^2	Hecto	h	10^{-15}	Femto	f
10	Deca	da	10^{-18}	Atto	a

A symbol for a unit should be provided with no more than one prefix. Example: 10^3 kg = 10^6 g = Mg (megagram).

Prefix and unit symbol are treated as a single entity. Example: 10^6 m^3 = $(10^2$ m$)^3$ = (hm)^3 = hm^3 (hectometer cubed).

Table A.4 Decimal fractions and decimal multiples of SI units that are given separate names

Physical quantity	Unit name	Symbol	Structure
Volume	Liter	l, L	$1\,l = 10^{-3}\ m^3 = 1\ dm^3$
Mass	Ton		
	(metric)	t	$1\,t = 10^3\ kg = 1\ Mg$
Pressure, mechanical stress	Bar	bar	$1\ bar = 10^5\ Pa = 10^5\ N/m^2$

Table A.5 Units of atomic physics that are defined independently of the fundamental SI units

Name	Symbol	Structure
Atomic mass unit	u	$1\ u \approx 1.6605655 \times 10^{-27}\ kg$
Electron volt	eV	$1\ eV \approx 1.6021892 \times 10^{-19}\ J$
Astronomic unit	(AU)	$1\ AU = 149597.870 \times 10^6\ m$
Parsec	pc	$1\ pc = 30857 \times 10^{12}\ m$

References: Le Système International d'Unités (SI). Bureau International des Poids et Mesures. 2d ed., Paris, 1973. Also, *Official Journal of the European Economic Community*, No. L262, pp. 204–216, September 27, 1976.

CONVERSION OF UNITS

The following conversion equations serve to translate older units into SI units. A boldface last digit indicates that the respective numerical factor is exact, i.e., fixed by definition.

1. Pressure

Bar	$1 \text{ bar} = 10^5 \text{ Pa} = 10^5 \text{ N/m}^2$
Physical atmosphere	$1 \text{ atm} = 760 \text{ Torr} = 101\ 325 \text{ Pa}$
Engineering (technical) atmosphere	$1 \text{ at} = 1 \text{ kp/cm}^2 = 98066.5 \text{ Pa}$
Meter of water (column)	$1 \text{ mH}_2\text{O} = 9806.65 \text{ Pa}$
Millimeter mercury (column)	$1 \text{ mm Hg} = 133.322 \text{ Pa}$
Pound force per square inch (also psi)	$1 \text{ lb f/in}^2 = 1 \text{ psi} = 6894.76 \text{ Pa}$
One inch water (column)	$1 \text{ in H}_2\text{O} = 249.089 \text{ Pa}$
One inch of mercury (column)	$1 \text{ in Hg} = 3386.39 \text{ Pa}$

2. Temperature

The following relations apply to the numerical values T_K, T_R, t_C, and t_F on the Kelvin scale, Rankine scale, Celsius scale, and Fahrenheit scale, respectively.

$$T_K = 273.15 + t_C = (5/9)T_R$$
$$T_R = 459.67 + t_F = 1.8\ T_K$$
$$t_C = (5/9)(t_F - 32) = T_K - 273.15$$
$$t_F = 1.8\ t_C + 32 = T_R - 459.67$$

3. Energy, work, heat

Erg	$1 \text{ erg} = 1 \text{ dyn} \times \text{cm} = 10^{-7}$ J
IT calorie	$1 \text{ cal}_{IT} = 4.1868$ J
Calorie (15°)	$1 \text{ cal}_{15} = 4.1855$ J
Thermochemical calorie	$1 \text{ cal}_{th} = 4.184$ J
Kilopond-meter	$1 \text{ kpm} = 9.80665$ J
Kilowatt-hour	$1 \text{ kWh} = 3.6 \times 10^6$ J
British thermal unit[†]	$1 \text{ Btu} = 1055.06$ J
Foot pound force	$1 \text{ ft lb f} = 1.35582$ J

4. Power or work rate, heat-flow rate

Erg per second	$1 \text{ erg/s} = 10^{-7}$ W
Calorie per second	$1 \text{ cal}_{IT}/\text{s} = 4.1868$ W
Kilocalorie per hour	$1 \text{ kcal}_{IT}/\text{h} = 1.163$ W
Kilopond-meter per second	$1 \text{ kpm/s} = 9.80665$ W
European horsepower	$1 \text{ PS} = 1 \text{ CV} = 7.3549875$ W
Horsepower	$1 \text{ hp} = 745.7$ W
British thermal unit per hour	$1 \text{ Btu/h} = 0.293971$ W
Foot pound force per second	$1 \text{ ft lb f/s} = 1.35582$ W

5. Specific heat

$1 \text{ kcal}_{IT}/\text{kg K} = 1 \text{ cal}_{IT}/\text{g K} = 4186.8$ J/kg K
$1 \text{ kcal}_{th}/\text{kg K} = 1 \text{ cal}_{th}/\text{g K} = 4184$ J/kg K
$1 \text{ Btu/lb R} = 4186.8$ J/kg K

6. Thermal conductivity

$1 \text{ kcal}_{IT}/\text{m h K} = 1.163$ W/m K $1 \text{ Btu/ft s R} = 6239.64$ W/m K
$1 \text{ cal}_{IT}/\text{cm s K} = 418.68$ W/m K

7. Coefficient of heat transfer, film coefficient

$1 \text{ kcal}_{IT}/\text{m}^2 \text{ h K} = 1.163$ W/m^2 K $1 \text{ Btu/ft}^2 \text{ s R} = 204{,}417$ W/m^2 K
$1 \text{ cal}_{IT}/\text{cm}^2 \text{ s K} = 41{,}868$ W/m^2 K

8. Thermal diffusivity, kinematic viscosity, coefficient of diffusion

$1 \text{ m}^2/\text{h} = 2.77777 \ldots \times 10^{-4} \text{ m}^2/\text{s}$ $1 \text{ ft}^2/\text{h} = 25.8064 \times 10^{-6} \text{ m}^2/\text{s}$
$1 \text{ ft}^2/\text{s} = 92.9030 \times 10^{-3} \text{ m}^2/\text{s}$

The unit of kinematic viscosity cm^2/s is also known as Stokes.

[†]This is the International Steam Tables British thermal unit defined as $1 \text{ kcal}_{IT}/\text{kg} = 1.8$ Btu/lb.

THE UNIVERSAL CONSTANTS OF PHYSICS

Quantity	Symbol and numerical value	Uncertainty, ppm
Speed of light in vacuum	$c = 299\ 792\ 458\ (1.2)$ m/s	0.004
Elementary charge	$e = 1.602\ 1892(46) \times 10^{-19}$ A s	2.9
Planck's constant	$h = 6.626\ 176(36) \times 10^{-34}$ J s	5.4
Avogadro's constant	$N_A = 6.022\ 045(31) \times 10^{23}$ mol^{-1}	5.1
Faraday's constant	$F = eN_A = 9.648\ 456(27) \times 10^4$ A s/mol	2.8
Universal gas constant	$R = 8.314\ 41(26)$ J/mol K	31
Standard temperature	$T_S = 273.15$ K	–
Standard pressure	$p_S = 101\ 325$ Pa	–
Molar volume under standard conditions	$V_{ms} = RT_S/p_S = 22.41383(70)$ m^3/kmol	31
Boltzmann's constant	$k = R/N_A = 1.380662(44) \times 10^{-23}$ J/K	32
Stefan-Boltzmann constant	$\sigma = 2\pi^5 k^4/15h^3 c^2$	
	$= 5.67032(71) \times 10^{-8}$ W/m^2 K^4	125
First radiation constant	$c_1 = 2\pi hc^2 = 3.741832(20) \times 10^{-16}$ Wm^2	5.4
Second radiation constant	$c_2 = hc/k = 0.014\ 387\ 86(45)$ mK	31
Gravitational constant	$G = 6.6720(41) \times 10^{-11}$ Nm^2/kg^2	615
Standard acceleration due to gravity	$g_S = 9.80665$ m/s^2	–

The numbers enclosed in parentheses represent the standard deviation of the corresponding value, expressed in terms of its last digits. Example: $R = 8.314\ 41(26)$ J/mol K $= (8.31441 \pm 0.00026)$ J/mol K.

Reference: CODATA Bulletin 11. Recommended Consistent Values of the Fundamental Physical Constants, 1973.

CHARACTERISTIC DIMENSIONLESS NUMBERS IN HEAT AND MASS TRANSFER

Name	Symbol	Definition[†]
Biot number	Bi	hL/k_S
Fourier number	Fo	at/L^2
Froude number	Fr	w^2/gL
Grashof number	Gr	$g\beta TL^3/\nu^2$
Lewis number	Le	$a/D = \text{Sc}/\text{Pr}$
Mach number	Ma	w/a^*
Nusselt number	Nu	hL/k_{fl}
Nusselt number of second kind	Nu*	h^*L/D
Péclet number	Pe	$wL/a = \text{Re} \times \text{Pr}$
Prandtl number	Pr	$\nu/a = \text{Sc}/\text{Le}$
Rayleigh number	Ra	$g\beta TL^3/\nu h$
Reynolds number	Re	wL/ν
Schmidt number	Sc	$\nu/D = \text{Le Pr}$
Stanton number	St	$h/w\rho c_p = \text{Nu}/\text{Re Pr}$
Stanton number of second kind	St*	$h^*/w = \text{Nu}^*/\text{Re Sc}$
Stefan number	Sn	$l/c_S(T_S - T_O)$
Strouhal number	Sr	Lf/w
Weber number	We	$w^3 L\rho/\sigma$

Note: The listing includes dimensionless numbers that occur in convective heat and mass transfer.

[†] a, thermal diffusivity; a^*, speed of sound; c_p, isobaric specific heat; c_S, specific heat of solid; D, coefficient of diffusion; f, frequency; g, local acceleration due to gravity; h, coefficient of heat transfer (film coefficient); h^*, coefficient of mass transfer; k_S or k_{fl}, thermal conductivity of solid or fluid, respectively; l, enthalpy of phase transition (latent heat); L, characteristic length; T, temperature difference; T_S, temperature of solidification; T_0, reference temperature; t, time; w, velocity; β, coefficient of thermal expansion; ν, kinematic viscosity; ρ, density; and σ, surface tension.

THERMOPHYSICAL PROPERTIES OF SOLIDS, LIQUIDS, AND GASES

Symbol	Quantity	Unit
T	Temperature	°C
ρ	Density	kg/m³
c_p	Isobaric specific heat	J/kg K
k	Thermal conductivity	W/m K
a	Thermal diffusivity $a = k/\rho c$	m² /s
b	Coefficient of heat penetration $b = (k\rho c_p)^{1/2}$	Ws$^{1/2}$ /m² K
μ	Dynamic viscosity	kg/ms = Pa s

I. Solids

	T, °C	ρ, kg/m³	c_p, J/kg K	k, W/m K	a, 10⁻⁶ m² /s	b, W s$^{1/2}$/ m² K
1. Metals and Alloys						
Aluminum 99.99	20	2,700	945	238	93.4	24,700
Antimony	0	6,690	209	230	164	17,900
Beryllium (sintered)	20	1,877	1,780	155	46.4	22,800
Bismuth	60	9,798	125	7.62	6.22	3,050

I. Solids (*Continued*)

	$T,$ °C	$\rho,$ kg/m³	$c_p,$ J/kg K	$k,$ W/m K	$a,$ 10^{-6} m²/s	$b,$ W s$^{1/2}$/ m² K
1. Metals and Alloys (*Continued*)						
Brass (58 Cu)	20	8,440	376	113	35.6	18,900
Brass (60 Cu)	20	8,400	376	113	35.8	18,900
Bronze (6 Sn, 9 Zn, 84 Cu, 1 Pb)	20	8,800	377	61.7	18.6	14,300
Cadmium	100	8,640	246	94.2	44.3	14,100
Chromium	20	6,900	457	69.1	21.9	14,800
Cobalt	20	8,780	427	69.1	18.4	16,100
Constantan (60 Cu, 40 Ni)	20	8,900	410	22.6	6.19	9,080
Copper (pure)	20	8,960	385	394	114	36,900
Copper (commercial)	20	8,300	419	372	107	36,000
Duraluminum (94-96 Al, 3-5 Cu, 0.5 Mg)	20	2,700	912	165	67.0	20,200
Gold (pure)	20	19,290	128	295	119	27,000
Iridium	20	22,400	133	147	49.3	20,900
Iron						
Cr-Ni steel (X 12 CrN; 18.8)	20	7,800	502	14.7	3.75	7,590
Cr-Al steel (X 10 CrAl 24), heat resistant	20	7,600	500	16.7	4.40	7,970
V 2 A steel (tempered) (0.2–0.6% C)	20	8,000	477	15.0	3.93	7,570
Cr-steel (X 8 Cr 17) stainless and acid resistant	20	7,700	460	25.1	7.09	9,430
Boiler plate, HIII	20	7,900	470	52.3	14.1	13,900
Lead	20	11,340	131	35.3	23.8	7,240
Magnesium	20	1,740	1,050	159	87.0	17,000
Manganese (α-Mn)	20	7,430	477	20.9	5.90	8,610
Manganin (84 Cu, 4 Ni, 12 Mn)	20	8,400	406	21.9	6.42	8,640
Molybdenum	20	10,200	272	147	53.0	20,200
Monel metal 505	60	8,360	544	19.7	4.33	9,460
Nickel (98.7% pure)	0	8,847	418	69.1	18.7	16,000
Niobium	20	8,570	267	52.3	22.9	10,900
Palladium	20	11,970	242	71.2	24.6	14,400
Platinum (very pure)	20	21,500	133	71.2	24.9	14,300
Potassium	20	860	766	196	298	11,400
Rhenium	20	21,020	138	48.1	16.6	11,800
Rhodium	20	12,500	246	151	49.1	21,500
Silver	20	10,497	234	408	166	31,700
Sodium	20	970	1,234	134	112	12,700
Tantalum	20	16,500	142	54.4	23.2	11,300
Tin (white β-tin)	20	7,290	221	62.8	39.0	10,100
Titanium	20	4,505	522	15.5	6.59	6,040
Tungsten	20	19,000	138	130	49.6	18,500
Uranium 238 (sintered 99.9)	500	18,000	174	30.3	9.67	9,740

I. Solids (*Continued*)

T, °C	ρ, kg/m³	c_p, J/kg K	k, W/m K	a, 10^{-6} m²/s	b, W s$^{1/2}$/ m² K

1. Metals and Alloys (*Continued*)

	T, °C	ρ, kg/m³	c_p, J/kg K	k, W/m K	a, 10^{-6} m²/s	b, W s$^{1/2}$/ m² K
UO₂	600	11,000	313	4.18	1.21	3,790
UO₂	1,000	10,960	326	3.05	0.854	3,300
UO₂	1,400	10,900	339	2.30	0.622	2,920
Vanadium	50	6,120	498	31.0	10.2	9,720
Wood's metal (50 Bi, 25 Pb, 12.5 Cd, 12.5 Sn)	20	1,056	147	12.8	82.5	1,410
Zinc	20	7,130	385	113	41.2	17,600
Zirconium	70	6,490	290	16.7	8.87	5,610

2. Structural Materials and Insulation

	T, °C	ρ, kg/m³	c_p, J/kg K	k, W/m K	a, 10^{-6} m²/s	b, W s$^{1/2}$/ m² K
Concrete (made with gravel, dry)	20	2,200	879	1.28	0.662	1,570
Mortar	20	1,900	800	0.93	0.61	1,190
Brickwork (dried in air)	20	1,400–1,800	840	0.58–0.81	0.49–0.54	830–1,100
Plaster	20	1,690	800	0.79	0.58	1,030
Asphalt	20	2,120	920	0.70	0.36	1,170
Cement (Portland, fresh, dry)	20	3,100	750	0.30	0.13	840
Gypsum	20	1,000	1,090	0.51	0.47	750
Beech (along fibers; 20% H₂O)	30	700	2,020	0.35	0.25	700
Oak (radially)	20	600–800	2,400	0.17–0.25	0.12–0.13	490–690
Fir (spruce), radially	20	410	2,700	0.14	0.13	390
Asbestos fiber	50	470	820	0.11	0.29	210
Slag wool	25	200	800	0.05	0.31	89
Mineral wool	50	200	920	0.046	0.25	92
Cork plates	30	190	1,880	0.041	0.11	120
Glass wool	0	200	660	0.037	0.28	70

3. Minerals and Glasses

	T, °C	ρ, kg/m³	c_p, J/kg K	k, W/m K	a, 10^{-6} m²/s	b, W s$^{1/2}$/ m² K
Earth (coarse-grained)	20	2,040	1,840	0.59	0.16	1,500
Fire clay	100	1,700–2,000	840	0.50–1.20	0.35–0.71	840–1,100
Clay	20	1,450	880	1.28	1.0	1,280
Quartz	20	2,100–2,500	780	1.40	0.72–0.85	1,500–1,650
Silica stone (85% SiC)	700	2,720	1,050	1.56	0.55	2,100
Slate (at right angles to lamination)	20	2,700	750	1.83	0.90	1,930
Sandstone	20	2,150–2,300	710	1.6–2.1	1.0–1.3	1,600–1,900

I. Solids (*Continued*)

	$T,$ °C	$\rho,$ kg/m³	$c_p,$ J/kg K	$k,$ W/m K	$a,$ 10^{-6} m²/s	$b,$ W s$^{1/2}$/ m² K
3. Minerals and Glasses (*Continued*)						
Chalkstone (CaCO₃, chalk)	20	2,000–3,000	740	2.2	1.0–1.5	1,800–2,200
Marble	20	2,500–2,700	810	2.8	1.3–1.4	2,400–2,500
Granite	20	2,750	890	2.9	1.2	2,700
Slate (parallel to lamination)	20	2,700	750	2.9	1.4	2,400
Rock salt	0	2,100–2,500	920	7.0	3.0–3.6	3,700–4,000
Mirror glass	20	2,700	800	0.76	0.35	1,280
Lead glass	20	2,890	680	0.70–0.93	0.36–0.47	1,170–1,350
Thermometer glass (Jena 16^{III})	20	2,580	780	0.97	0.48	1,400
Pyrex	20	2,240	774	1.06	0.61	1,360
Window glass	20	2,480	700–930	1.16	0.50–0.67	1,420–1,640
Quartz glass	20	2,210	730	1.40	0.87	1,500
4. Synthetic Materials						
Moldable amino resin	20	1,500	1,670	0.35	0.14	940
Polyethylene	20	920	2,300	0.35	0.165	860
Polyurethane	20	1,200	2,090	0.32	0.128	800
Polyamide	20	1,130	2,300	0.28–0.30	0.108–0.115	850–880
Polyvinyl carbazol	20	1,190	1,250	0.26	0.175	620
Bakelite	20	1,270	1,590	0.23	0.114	680
Synthetic rubber	20	1,150	1,970	0.23	0.101	720
Celluloid	20	1,380	1,670	0.23	0.10	730
Polytetrafluoroethylene (Teflon)	20	2,200	1,040	0.23	0.10	725
Rubber (soft)	20	1,100	1,670	0.16–0.23	0.087–0.125	540–650
Unsaturated polyester resin	20	1,100	1,750	0.185	0.096	600
Acrylic glass (Plexiglas)	20	1,180	1,440	0.184	0.108	560
Cast phenol resin	20	1,330	1,460	0.174	0.090	580
Hard rubber	20	1,150	1,420	0.160	0.098	510
Polyvinyl chloride (PVC)	20	1,380	960	0.15	0.113	445
Polystyrol	20	1,050	1,250	0.14	0.107	430
Moldable anilin resin	20	1,200	1,670	0.12	0.060	490
Foam rubber	20	400–500	1,670	0.070–0.092	0.105–0.110	220–280
Polystyrol foam	20	15–100	1,250	0.029–0.045	0.36–1.54	23–75

I. Solids (*Continued*)

	$T,$ °C	$\rho,$ kg/m³	$c_p,$ J/kg K	$k,$ W/m K	$a,$ 10^{-6} m²/s	$b,$ W s$^{1/2}$/ m² K
5. Various Materials						
Wool	20	100	1,720	0.036	0.21	79
Artificial silk	35	100	1,330	0.049	0.37	81
Cotton	30	81	1,150	0.059	0.63	74
Bark	25	340	1,260	0.074	0.17	180
Coal dust	30	730	1,300	0.12	0.13	340
Paper (regular)	20	700	1,200	0.12	0.14	320
Leather (dry)	20	860	1,500	0.12– 0.15	0.09– 0.12	390– 440
Fat	20	910	1,930	0.17	0.097	550
Coal	20	1,200– 1,500	1,260	0.26	0.14– 0.17	630– 700
Sulfur	20	2,070	720	0.27	0.18	630
Paraffin	30	870– 925	2,900	0.24– 0.27	0.095– 0.10	780– 850
Snow (firm)	0	560	2,100	0.46	0.39	740
Ice	0	917	2,040	2.25	1.20	2,050
Slag	20	2,500– 3,000	840	0.57	0.23– 0.27	1,100– 1,200
Porcelain	95	2,400	1,080	1.03	0.40	1,600
Clay	20	1,450	880	1.3	1.0	1,300
Sugar (fine)	0	1,600	1,250	0.58	0.29	1,080
Carborundum (SiC)	100	1,500	620	58	62	7,300
Graphite (firm, natural)	20	2,000– 2,500	610	155	1.02– 1.27	13,700– 15,400

II. Liquids

	$T,$ °C	$\rho,$ kg/m³	$c_p,$ J/kg K	$k,$ W/m K	$a,$ 10^{-6} m²/s	$b,$ W s$^{1/2}$/ m² K	$\eta,$ 10^{-3} kg/sm
1. Various Liquids							
Ammonia (NH_3)	20	610	4,770	0.49	0.17	1,200	0.22
Acetone (C_3H_6O)	20	790	2,210	0.18	0.10	560	0.32
Chloroform ($CHCl_3$)	20	1,490	980	0.12	0.082	420	0.57
Carbon tetrachloride (CCl_4)	20	1,595	840	0.105	0.078	375	0.97
Water	20	998	4,183	0.598	0.143	1,580	1.00
Toluene ($C_6H_5CH_3$)	20	866	1,700	0.14	0.095	450	0.59

II. Liquids (*Continued*)

	$T,$ °C	$\rho,$ kg/m³	$c_p,$ J/kg K	$k,$ W/m K	$a,$ 10^{-6} m²/s	$b,$ W s$^{1/2}$/ m² K	$n,$ 10^{-3} kg/sm
1. Various Liquids (*Continued*)							
Benzene	20	879	1,740	0.154	0.10	485	0.65
Ethyl alcohol (C_2H_5OH)	20	789	2,430	0.18	0.094	590	1.20
Turpentine	0	860	1,720	0.14	0.095	460	1.50
Ethylene glycol							
[$C_2H_4(OH)_2$]	100	1,056	2,740	0.26	0.090	870	2.40
Glycerine [$C_3H_5(OH)_3$]	20	1,260	2,430	0.27	0.088	910	15
Sulfuric acid (H_2SO_4)	10	1,830	1,410	0.54	0.21	1,180	27
Diphyl (73.5% diphenyl							
oxide, 26.5%							
diphenyl)	20	1,060	1,590	0.14	0.083	490	5.8
Brine (20% $MgCl_2$)	−20	1,180	3,000	0.39	0.11	1,200	13
Freons (at saturation pressure)							
Freon 11 ($CFCl_3$)	20	1,530	879	0.10	0.074	370	0.54
Freon 12 (CF_2Cl_2)	20	1,390	934	0.084	0.065	330	0.29
Freon 13 (CF_3Cl)	20	1,120	1,210	0.058	0.043	280	0.22
Freon 22 (CHF_2Cl)	0	1,290	1,180	0.098	0.064	390	0.24
Freon 114							
($CF_2Cl \cdot CF_2Cl$)	0	1,530	940	0.076	0.053	330	0.47
Transformer oil	60	842	2,090	0.12	0.068	460	7.3
Lubricating oil	60	845	2,020	0.14	0.082	490	4.2
Olive oil	60	920	1,970	0.16	0.088	540	81
Aero-engine oil	60	868	2,010	0.14	0.080	490	71
Silicone oil	20	970	1,470	0.17	0.12	490	240
2. Liquid Metals and Alloys							
Potassium	200	795	791	46	73	5,400	0.34
Rubidium	39	1,480	382	22	39	3,500	0.48
Cesium	28	1,840	251	13.6	29	2,500	0.48
Sodium	100	927	1,390	86	67	10,500	0.71
Tin	300	6,940	255	32	18	7,500	1.67
Bismuth	300	10,000	150	14.6	9.7	4,700	1.66
Na-K alloy (22 Na)	100	847	941	23	29	4,300	0.53
Lead	400	10,600	147	15.1	9.7	4,900	2.10
Mercury	20	13,600	139	8.0	4.2	3,900	1.55
Lead-bismuth alloy							
(44.5% Pb)	200	10,500	146	11.7	7.6	4,200	2.50
Lithium	200	515	4,140	46	22	9,900	0.57

III. Gases ($p = 1.01325$ bar)

	$T,$ °C	$\rho,$ kg/m³	$c_p,$ J/kg K	$k,$ W/m K	$a,$ 10^{-3} m²/s	$\eta,$ 10^{-6} kg/s m
1. Pure Inorganic Gases (vapors)						
Helium (He)	0	0.18	5,200	0.143	153	19
Neon (Ne)	0	0.90	1,030	0.046	50	30
Argon (Ar)	0	1.78	524	0.018	19.2	21
Hydrogen (H_2)	0	0.09	14,050	0.171	13.6	8.4
Oxygen (O_2)	0	1.43	909	0.024	18.4	19.2
Nitrogen (N_2)	0	1.25	1,038	0.024	18.5	16.6
Air	0	1.29	1,005	0.024	18.5	17.2
Chlorine (Cl_2)	0	3.17	473	0.0081	5.4	12.3
Carbon monoxide (CO)	0	1.25	1,038	0.023	17.7	16.6
Carbon dioxide (CO_2)	0	1.96	816	0.015	9.3	13.7
Nitric oxide (NO)	0	1.34	971	0.024	18.4	18
Sulfur dioxide (SO_2)	0	2.86	586	0.0086	5.0	12
Steam (H_2O)	100	0.598	2,028	0.025	20.6	12
Ammonia (NH_3)	0	0.77	2,056	0.022	14	9.3
2. Pure Organic Gases (vapors)						
Methane (CH_4)	0	0.72	2,165	0.030	19.2	10
Ethane (C_2H_6)	0	1.35	1,650	0.018	8.1	8.6
Propane (C_3H_8)	0	2.01	1,550	0.015	4.8	7.5
Ethylene (C_2H_4)	0	1.26	1,460	0.017	9.2	9.4
Acetylene (C_2H_2)	0	1.17	1,616	0.018	9.5	9.6
Freons (at saturation pressure)						
Freon 11 ($CFCl_3$)	0	2.48	549	0.0077	5.7	8.5
Freon 12 (CF_2Cl_2)	0	17.7	547	0.0097	1.0	9.9
Freon 13 (CF_3Cl)	0	134	620	0.0115	0.14	13
Freon 22 (CHF_2Cl)	0	21.5	636	0.0107	0.8	9.6
Freon 114 ($CF_2Cl\cdot CF_2Cl$)	20	9.6	653	0.0109	1.7	11

THERMAL CONDUCTIVITY OF LIQUIDS
AT MODERATE PRESSURES

$k = k_0 - k'T$, where T is temperature on Celsius scale[a]

Liquid	k_0, W/m K	10^4 k', W/m K^2	Temperature range, °C
	Refrigerants[b]		
R10	0.1068	2.19	−20–105
R11	0.0945	2.81	−105–75
R12	0.0783	3.66	−120–25
R13	0.0497	5.22	−125–0
R14	−0.0008	7.92	−125−−70
R20	0.1217	2.81	−55–75
R21	0.1118	3.81	−125–50
R22	0.1001	4.95	−125–30
R23	0.0760	7.48	−125−−5
R30	0.1469	3.99	−95–50
R31	0.1464	5.52	−100–25
R32	0.1474	8.02	−125–25
R112	0.0860	1.65	30–100
R113	0.0802	2.05	−30–70
R114	0.0710	2.61	−90–30
R114B2	0.0667	1.58	−105–50
R115	0.0603	3.24	−100–0

(*See footnotes on p. 179.*)

$k = k_0 - k'T$, where T is temperature on Celsius scale[a] (Continued)

Liquid	k_0, W/m K	10^4 k', W/m K²	Temperature range, °C
	Refrigerants[b] (Continued)		
R116	0.0459	4.14	−95−−5
R113a	0.0975	3.61	−100−50
R133aB1	0.0882	2.61	−90−50
1,2,4,5-Tetrachlorbenzene	0.1250	1.62	140−185
Hexachlorbenzole	0.1201	1.52	230−250
Toluol	0.1406	2.80	−20−110
o-Xylene	0.1378	2.27	−20−80
m-Xylene	0.1393	2.39	−40−80
p-Xylene	0.1354	2.41	15−80
1,2,3-Trimethylbenzene	0.1333	1.64	−5−80
1,2,4-Trimethylbenzene	0.1347	1.97	−50−80
1,3,5-Trimethylbenzene	0.1429	2.39	−50−80
1,2,4,5-Tetramethylbenzene	0.1342	1.15	100−130
Pentamethylbenzene	0.1294	1.03	60−150
Hexamethylbenzene	0.1098	0.14	170−240
Ethylbenzene	0.1375	2.41	−80−80
u-Propylbenzene	0.1347	1.85	−85−80
u-Butylbenzene	0.1343	1.72	−70−100
u-Penthylbenzene	0.1337	1.46	−70−100
u-Hexylbenzene	0.1335	1.48	−55−100
Naphthalene	1.1374	0.86	85−130
Anthracene	0.1431	0.80	225−275
Phenanthrene	0.1325	0.32	105−210
Pyrene	0.1294	0.41	155−235
R152a	0.1165	4.97	−110−25
R214	0.0782	1.43	−90−100
R215	0.0739	1.72	−75−75
R216	0.0675	2.09	−125−25
RC318	0.0737	3.38	−40−25
RC51-12	0.0631	1.57	−20−50
R846	0.0648	3.80	−45−0
R500	0.0846	3.94	−120−20
R502	0.0742	3.91	−125−25
R503	0.0569	5.80	−125−−5
R504	0.0928	5.08	−75−25
	Benzene derivatives[c]		
Benzene	0.1522	3.23	10−75
Monofluorobenzene	0.1329	2.80	−35−80
Monochlorobenzene	0.1320	2.27	−40−80

(See footnotes on p. 179.)

$k = k_0 - k'T$, where T is temperature on Celsius scale[a] (Continued)

Liquid	k_0, W/m K	$10^4 \ k'$, W/m K^2	Temperature range, °C
Benzene derivatives[c] (Continued)			
Monobromobenzene	0.1143	1.63	−25–80
Monoiodobenzene	0.1010	0.95	−25–80
o-Dichlorobenzene	0.1246	1.64	−10–80
m-Dichlorobenzene	0.1201	1.51	−20–80
p-Dichlorobenzene	0.1172	1.37	60–90
1,2,3-Trichlorobenzene	0.1138	0.75	60–105
1,2,4-Trichlorobenzene	0.1152	1.33	20–80
1,3,5-Trichlorobenzene	0.1211	1.75	70–100
1,2,3,4-Tetrachlorobenzene	0.1081	0.83	50–130
Alcohols (water content in % by mass)[d]			
Methanol (0.1)	0.2040	3.08	10–40
Ethanol (0.2)	0.1712	3.02	10–30
u-Propanol (0.05)	0.1576	2.31	10–40
Isopropanol (0.1)	0.1395	2.02	10–40
u-Butanol (0.1)	0.1534	2.11	10–55
sec-Butanol (0.05)	0.1400	2.03	10–55
Isobutanol (0.05)	0.1353	1.66	10–55
tert-Butanol (0.1)	0.1110	1.27	30–50
Dialkylphthalates[e]			
Dimethylphthalate	0.1507	0.97	10–85
Diethylphthalate	0.1456	1.15	10–85
Diisopropylphthalate	0.1311	1.01	10–85
Di-u-butylphthalate	0.1387	1.12	10–85
Diisobutylphthalate	0.1280	1.00	10–85
Diethylhexylphthalate	0.1369	0.97	10–85

[a]Example: The thermal conductivity of toluol at 50°C is $k_{50} = (0.1406 − 0.000280 \times 50)$ W/m K = 0.1266 W/m K.

[b]Tauscher, W., Wärme-Stoffübertrag., vol. 1, pp. 140–146, 1968.

[c]Bachmann, R., Wärme-Stoffübertrag., vol. 2, pp. 129–134, 1969.

[d]Poltz, H. and Jugel, R., Wärme-Stoffübertrag., vol. 1, pp. 197–201, 1968. Due to the effects of radiation absorption, k depends weakly on the layer thickness. The values indicated apply to layers 1-mm thick, approximately.

[e]Poltz, H., Wärme-Stoffübertrag., vol. 3, pp. 247–250, 1970.

TABLES OF MATHEMATICAL FUNCTIONS

Table G.1 Gauss's error function and integral complementary error function normalized with the factor $\pi^{1/2}$

x	erf (x)	$\sqrt{\pi}$ ierfc (x)	x	erf (x)	$\sqrt{\pi}$ ierfc (x)
0.0	0.00000	1.00000	1.6	0.97635	0.01023
0.1	0.11246	0.83274	1.7	0.98379	0.00673
0.2	0.22270	0.68524	1.8	0.98909	0.00436
0.3	0.32863	0.55694	1.9	0.99279	0.00277
0.4	0.42839	0.44688	2.0	0.99532	0.00173
0.5	0.52050	0.35385	2.1	0.99702	0.00107
0.6	0.60386	0.27639	2.2	0.99814	0.00064
0.7	0.67780	0.21287	2.3	0.99886	0.00038
0.8	0.74210	0.16160	2.4	0.99931	0.00022
0.9	0.79691	0.12088	2.5	0.99959	0.00013
1.0	0.84270	0.08907	2.6	0.99976	0.00007
1.1	0.88021	0.06463	2.7	0.99987	0.00004
1.2	0.91031	0.04617	2.8	0.99992	0.00002
1.3	0.93401	0.03246	2.9	0.99996	0.00001
1.4	0.95229	0.02246	3.0	0.99998	0.00001
1.5	0.96611	0.01528			

Table G.2 Exponential integral

x	$-Ei(-x)$	x	$-Ei(-x)$
0.00	$+\infty$		
0.01	4.0379	1.60	0.08631
0.02	3.3547	1.70	0.07465
0.04	2.6813	1.80	0.06471
0.07	2.1508	1.90	0.05620
0.10	1.8229	2.00	0.04890
0.15	1.4645	2.20	0.03719
0.20	1.2227	2.40	0.02844
0.30	0.9057	2.60	0.02185
0.40	0.7024	2.80	0.01686
0.50	0.5598	3.00	0.01305
0.60	0.4544	3.20	0.01013
0.70	0.3738	3.40	0.00789
0.80	0.3106	3.60	0.00616
0.90	0.2602	3.80	0.00482
1.00	0.2194	4.00	0.00378
1.10	0.18599	4.20	0.00297
1.20	0.15841	4.40	0.00234
1.30	0.13545	4.60	0.00184
1.40	0.11622	4.80	0.00145
1.50	0.10002	5.00	0.00115

Table G.3 Modified Bessel function of the second kind

x	$K_0(x)$	x	$K_0(x)$
0.0	$+\infty$		
0.1	2.4271	1.6	0.18795
0.2	1.7527	1.7	0.16550
0.3	1.3724	1.8	0.14593
0.4	1.1145	1.9	0.12885
0.5	0.9244	2.0	0.11389
0.6	0.7775	2.2	0.08927
0.7	0.6605	2.4	0.07022
0.8	0.5653	2.6	0.05540
0.9	0.4867	2.8	0.04382
1.0	0.4210	3.0	0.03474
1.1	0.3656	3.2	0.02759
1.2	0.3185	3.5	0.01960
1.3	0.2782	4.0	0.01116
1.4	0.2437	4.5	0.00640
1.5	0.2138	5.0	0.00369

INDEX